Howard Rheingold

The History

and

Future of

Mind–Expanding

Technology

TOOLS FOR THOUGHT

The MIT Press   Cambridge, Massachusetts   London, England

First MIT Press edition 2000

Library of Congress Cataloging-in-Publication Data

Rheingold, Howard.
  Tools for thought : the history and future of mind-expanding technology / Howard Rheingold.
    p. cm.
  Reprint. Originally published: New York : Simon & Schuster, 1985.
  Includes bibliographical references and index.
  ISBN 0-262-68115-3 (pbk. : alk. paper)
  1. Microcomputers—History. 2. Technological innovations—History. I. Title.
QA76.5. R467 2000
303.48'34—dc21   99-087051

Preface to the MIT Press Edition

At the beginning of the 1980s, powerful personal computers and global computer networks of millions of computers were not the fact of life they became at the end of the 1990s. Indeed, the world of information and communication technology that influences so much of our lives today was not created by the existing computer industry, nor was it championed by the orthodoxy of computer science. Rather, it was built by a handful of rebels who weren't seeking fame or fortune, but spent their lives creating a new tool for enhancing human thought. They created it because they wanted it for their personal use, because it was a cool thing to do, and because they thought it would improve the human lot.

When Apple and Microsoft were fledgling companies, most of what was written about the emerging personal computer industry was about teenage millionaires. If you were to trust what you read in the popular press, the personal computer was invented by Steve Jobs and Bill Gates. In fact, any journalist could get a pretty good idea of what life might be like two decades in the future if they visited Xerox Palo Alto Research Center in the late 1970s—the place Jobs and Gates got all their best ideas. The real story of where PCs and networks came from was, to me, both more interesting and more fundamentally important than the popular mythology of Silicon Valley. Seventeen years ago, I sought out PARC's Alan Kay, ARPA's J. C. R Licklider and Bob Taylor, and SRI's Doug Engelbart, because I was personally fascinated with the idea that computers could one day be used to help people think, communicate, and solve problems together.

In 1999, I revisited Engelbart, Kay, Taylor, and others I first encountered in 1983. We talked about the way the future looked when they started creating it, the way it has actually turned out, and the possible futures for mind-amplifying technology. Retrospective futurism about technology is considerably easier than trying to foresee where our tools are really going to take us next. It's particularly tasty if MIT Press pays you to do it. At the end of this edition is an afterword that attempts to look back on how we looked forward in 1983, when the book was written.

Acknowledgments

This book would not have been conceived and could not have been written without the generous and patient assistance of many people. My heartfelt thanks to Rita Aero, Avron Barr, John Brockman, Donald Day, Robert Eckhardt, Doug Engelbart, Brenda Laurel, Howard Levine, Judith Maas, Geraldine Rheingold, Alan Rinzler, Charles Silver, Marshall Smith, Bob Taylor, David Rodman, and Gloria Warner.

To Nathan Rheingold,
who gave me the most important thing a man
can ever give his son—an example.

Contents

TOOLS FOR THOUGHT

CHAPTER ONE

The Computer Revolution Hasn't Happened Yet

South of San Francisco and north of Silicon Valley, near the place where the pines on the horizon give way to live oaks and radiotelescopes, an unlikely subculture has been creating a new medium for human thought. When the mass-production models of present prototypes reach our homes, offices, and schools, our lives are going to change dramatically.

The first of these mind-amplifying machines will be descendants of the devices now known as personal computers, but they will resemble today's information processing technology no more than a television resembles a fifteenth-century printing press. They aren't available yet, but they will be here soon. Before today's first-graders graduate from high school, hundreds of millions of people around the world will join together to create new kinds of human communities, making use of a tool that a small number of thinkers and tinkerers dreamed into being over the course of the past century.

Nobody knows whether this will turn out to be the best or the worst thing the human race has done to itself, because the outcome of this empowerment will depend in large part on how we react to it and what we choose to do with it. The human mind is not going to be replaced by a machine, at least not in the foreseeable future, but there

13

is little doubt that the worldwide availability of fantasy amplifiers, intellectual toolkits, and interactive electronic communities will change the way people think, learn, and communicate.

It looks as if this latest technology-triggered transformation of society could have even more intense impact than the last time human thought was augmented, five hundred years ago, when the Western world learned to read. Less than a century after the invention of movable type, the literate community in Europe had grown from a privileged minority to a substantial portion of the population. People's lives changed radically and rapidly, not because of printing machinery, but because of what that invention made it possible for people to know. Books were just the vehicles by which ideas escaped from the private libraries of the elite and circulated among the population.

The true value of books emerged from the community they made possible, an intellectual community that is still alive all over the world. The printed page has been a medium for the propagation of ideas about chemistry and poetry, evolution and revolution, democracy and psychology, technology and industry, and many other notions far beyond the ken of the people who invented movable type and started cranking out Bibles.

Because mass production of sophisticated electronic devices can lag ten years or more behind the state of the art in research prototypes, the first effects of the astonishing achievements in computer science since 1960 have only recently begun to enter our lives. Word processors, video games, educational software, and computer graphics were unknown terms to most people only ten years ago, but today they are the names for billion-dollar industries. And the experts agree that the most startling developments are yet to come.

A few of the pioneers of personal computing who still work in the computer industry can remember the birth of the dream, when the notion of personal computing was an obscure heresy in the ranks of the programming priesthood. Thirty years ago, the overwhelming majority of the people who designed, manufactured, programmed, and used computers subscribed to a single idea about the proper (and possible) place of computers in society: "Computers are mysterious devices meant to be used for mathematical calculations." Period. Computer technology was believed to be too fragile, valuable, and complicated for nonspecialists.

In 1950, you could count the people who took exception to this dogma on the fingers of one hand. The dissenting point of view shared by those few people involved a different way of thinking about how computers might be used. The dissenters shared a vision of *personal* computing in which computers would be used to enhance the most creative aspects of human intelligence—for everybody, not just the technocognoscenti.

Those who questioned the dogma of data processing agreed that computers can help us calculate, but they also suspected that if the devices could be made more interactive, these tools might help us to speculate, build and study models, choose between alternatives, and search for meaningful patterns in collections of information. They wondered whether this newborn device might become a communication medium as well as a calculating machine.

These heretical computer theorists proposed that if human knowledge is indeed power, then a device that can help us transform information into knowledge should be the basis for a very powerful technology. While most scientists and engineers remained in awe of giant adding machines, this minority insisted on thinking about how computers might be used to assist the operation of human minds in nonmathematical ways.

Tools for Thought focuses on the ideas of a few of the people who have been instrumental in creating yesterday's, today's, and tomorrow's human-computer technology. Several key figures in the history of computation lived and died centuries or decades ago. I call these people, renowned in scientific circles but less well known to the public, the *patriarchs*. Other cocreators of personal computer technology are still at work today, continuing to explore the frontiers of mind-machine interaction. I call them the *pioneers*.

The youngest generation, the ones who are exploring the cognitive domains we will all soon experience, I call the *infonauts*. It is too early to tell what history will think of the newer ideas, but we're going to take a look at some of the things the latest inner-space explorers are thinking, in hopes of catching some clues to what (and how) everybody will be thinking in the near future.

As we shall see, the further limits of this technology are not in the hardware, but in our minds. The digital computer is based on a theoretical discovery known as "the universal machine," which is not actually a tangible device but a mathematical description of a machine capable of simulating the actions of any other machine. Once you have created a general-purpose machine that can imitate any other machine, the future development of the tool depends only on what tasks you can think to do with it. For the immediate future, the issue of whether machines can become intelligent is less important than learning to deal with a device that can become whatever we clearly imagine it to be.

The pivotal difference between today's personal computers and tomorrow's intelligent devices will have less to do with their hardware than their *software*—the instructions people create to control the operations of computing machinery. A program is what tells the general-purpose machine to imitate a specific kind of machine. Just as the hardware basis for computing has evolved from relays to vacuum tubes to transistors to integrated circuits, the programs have evolved, as well. When information processing grows into knowledge processing, the true personal computer will reach beyond hardware and connect with a vaster source of power than that of electronic microcircuitry—the power of human minds working in concert.

The nature of the world we create in the closing years of the twentieth century will be determined to a significant degree by our attitudes toward this new category of tool. Many of us who were educated in the precomputer era shall be learning new skills. The college class of 1999 is already on its way. It is important that we realize today that those skills of tomorrow will have little to do with how to operate computers and a great deal to do with how to use augmented intellects, enhanced communications, and amplified imaginations.

Forget about "computer literacy" or obfuscating technical jargon, for these aberrations will disappear when the machines and their programs grow more intelligent. The reason for building a personal computer in the first place was to enable people to do what people do best by using machines to do what machines do best. Many people are afraid of today's computers because they have been told that these machines are smarter than they are—a deception that is reinforced by the rituals that novices have been forced to undergo in order to use computers. In fact, the burden of communication should be on the machine. A computer that is difficult to use is a computer that's too dumb to understand what you want.

If the predictions of some of the people in this book continue to be accurate, our whole environment will suddenly take on a kind of intelligence of its own sometime between now and the turn of the century. Fifteen years from now there will be a microchip in your telephone receiver with more computing power than all the technology the Defense Department can buy today. All the written knowledge in the world will be one of the items to be found in every schoolchild's pocket.

The computer of the twenty-first century will be everywhere, for better or worse, and a more appropriate prophet than Orwell for this eventuality might well be Marshall McLuhan. If McLuhan was right about the medium being the message, what will it mean when the entire environment becomes the medium? If such a development does occur as predicted, it will probably turn out differently from even the wildest "computerized household" scenarios of the recent past.

The possibility of accurately predicting the social impact of any new technology is questionable, to say the least. At the beginning of the twentieth century, it was impossible for average people or even the most knowledgeable scientists to envision what life would be like for their grandchildren, who we now know would sit down in front of little boxes and watch events happening at that moment on the other side of the world.

Today, only a few people are thinking seriously about what to do with a living room wall that can tell you anything you want to know, simulate anything you want to see, connect you with any person or group of people you want to communicate with, and even help you find out what it is you want to know when you aren't entirely sure. In the 1990s, it might at last become feasible for people to "think as no human being has ever thought" and for computers to "process data in a way not approached by the information-handling machines we know today," as J.C.R. Licklider, one of the most influential pioneers, predicted in 1960, a quarter of a century before the hardware would begin to catch up with his ideas.

The earliest predictions about the impact of computing machinery occurred quite a bit earlier than 1960. The first electronic computers were invented by a few unusual individuals, who often worked alone, during World War II. Before the actual inventors of the 1940s were the software patriarchs of the 1840s. And before them, thousands of years ago, the efforts of thinkers from many different cultures to find better ways to use symbols as tools led to the invention of mathematics and logic. It was these formal systems for manipulating symbols that eventually led to computation. Links in what we can now see as a continuous chain of thought were created by a series of Greek philosophers, British logicians, Hungarian mathematicians, and American inventors.

Most of the patriarchs had little in common with each other, socially or intellectually, but in some ways they were very much alike. It isn't surprising that they were exceptionally intelligent, but what is unusual is that they all seem to have been preoccupied with the power of their own minds. For sheer intellectual adventure, many intelligent people pursue the secrets of the stars, the mysteries of life, the myriad ways to use knowledge to accomplish practical goals. But what the software ancestors sought to create were tools to amplify the power of their own brains—machines to take over what they saw as the more mechanical aspects of thought.

Perhaps as an occupational hazard of this dangerously self-reflective enterprise, or as a result of being extraordinary people in restrictive social environments, the personalities of these patriarchs (and matriarchs) of computation reveal a common streak of eccentricity, ranging from the mildly unorthodox to the downright strange:

• **Charles Babbage and Ada, Countess of Lovelace,** lived in the London of Dickens and Prince Albert (and knew them both). A hundred years before some of the best minds in the world used the resources of a nation to build a digital computer, these two eccentric inventor-mathematicians dreamed of building their "Analytical Engine." He constructed a partial prototype and she used it, with notorious lack of success, in a scheme to win a fortune at the horse races. Despite their apparent failures, Babbage was the first true computer designer, and Ada was history's first programmer.

• **George Boole** invented a mathematical tool for future computer-builders—an "algebra of logic" that was used nearly a hundred years later to link the process of human reason to the operations of machines. The idea came to him in a flash of inspiration when he was walking across a meadow one day, at the age of seventeen, but it took him twenty years to teach himself enough mathematics to write *The Laws of Thought.*

Although Boole's lifework was to translate his inspiration into an algebraic system, he continued to be so impressed with the suddenness and force of the revelation that hit him that day in the meadow that he also wrote extensively about the powers of the unconscious mind. After his death Boole's widow turned these ideas into a kind of human potential cult, a hundred years before the "me decade."

• **Alan Turing** solved one of the most crucial mathematical problems of the modern era at the age of twenty-four, creating the theoretical basis for computation in the process. Then he became the top code-breaker in the world—when he wasn't bicycling around wearing a gas mask or running twenty miles with an alarm clock tied around his waist. If it hadn't been for the success of Turing's top-secret wartime mission, the Allies might have lost World War II. After the war, he created the field of artificial intelligence and laid down the foundations of the art and science of programming.

He was notoriously disheveled, socially withdrawn, sometimes loud and abrasive, and even his friends thought he carried nonconformity to weird extremes. At the age of forty-two, he committed suicide, hounded cruelly by the same government he helped save.

• **John von Neumann** spoke five languages and knew dirty limericks in all of them. His colleagues, famous thinkers in their own right, all agreed that the operations of Johnny's mind were too deep and far too fast to be entirely human. He was one of history's most brilliant physicists, logicians, and mathematicians, as well as the software genius of the group who invented the first electronic digital computer.

John von Neumann was the center of the group who created the "stored-program" concept that made truly powerful computers possible, and he specified a template that is still used to design almost all computers—the "von Neumann architecture." When he died, the Secretaries of Defense, the Army, Air Force, and Navy and the Joint Chiefs of Staff were all gathered around his bed, attentive to his last gasps of technical and policy advice.

• **Norbert Wiener,** raised to be a prodigy, graduated from Tufts at fourteen, earned his Ph.D. from Harvard at eighteen, and studied with Bertrand Russell at nineteen. Wiener had a different kind of personality than his contemporary and colleague, von Neumann. Although involved in the early years of computers, he eventually refused to take part in research that could lead to the construction of weapons. Scarcely less brilliant than von Neumann, Wiener was vain, sometimes paranoid, and not known to be the life of the party, but he made important connections between computers, living organisms, and the fundamental laws of the physical universe. He guarded his ideas and feuded with other scientists, writing unpublished novels about mathematicians who did him wrong.

Wiener's conception of *cybernetics* was partially derived from "pure" scientific work in mathematics, biology, and neurophysiology, and partially derived from the grimly applied science of designing automatic antiaircraft guns. Cybernetics was about the nature of control and communication systems in animals, humans, and machines.

• **Claude Shannon,** another lone-wolf genius, is still known to his neighbors in Cambridge, Massachusetts, for his skill at riding a unicycle. In 1937, as a twenty-one-year-old graduate student, he showed that Boole's logical algebra was the perfect tool for analyzing the complex networks of switching circuits used in telephone systems and, later, in computers. During the war and afterward, Shannon established the mathematical foundation of *information theory*. Together with cybernetics, this collection of theorems about information and communication created a new way to understand people and machines—and established information as a cosmic fundamental, along with energy and matter.

The software patriarchs came from wildly different backgrounds. Then as now, computer geniuses were often regarded as "odd" by those around them, and their reasons for wanting to invent computing devices seem to have been as varied as their personalities. Something about the notion of a universal machine enticed mathematicians and philosophers, logicians and code-breakers, whiz kids and bomb-builders. Even today, the worlds of computer research and the software business bring together an unlikely mixture of entrepreneurs and evangelists, futurians and utopians, cultists, lunatics, obsessives, geniuses, pranksters, and fast-buck artists.

Despite their outward diversity, the computer patriarchs of a hundred years ago and the cyberneticians of the World War II era appear to have shared at least one characteristic with each other and with software pioneers and infonauts of more recent vintage. In recent years, the public has become more aware of a subculture that sprouted in Cambridge and Palo Alto and quietly spread through a national network of fluorescent-lit campus computer centers for the past two decades—the mostly young, mostly male, often brilliant, sometimes bizarre "hackers," or self-confessed compulsive programmers. Sociologists and psychologists of the 1980s are only beginning to speculate about the deeper motivations for this obsession, but any latter-day hacker will admit that the most fascinating thing in his life is his own mind, and tell you that he regards intense, prolonged interaction with a computer program as a particularly satisfying kind of dialogue with his own thoughts.

A little touch of the hacker mentality seems to have affected all of the major players in this story. From what we know today about the patriarchs and pioneers, they all appear to have pursued a vision of a new way to use their minds. Each of them was trying to create a mental lever. Each of them contributed indispensable components of the device that was eventually assembled. But none of them encompassed it all.

The history of computation became increasingly complex as it progressed from the patriarchs to the pioneers. At the beginning, many of the earliest computer scientists didn't know that their ideas would end up in a kind of machine. Almost all of them worked in isolation. Because of their isolation from one another, the common intellectual ancestors of the modern computer are relatively easy to discern in retrospect. But since the 1950s, with the proliferation of researchers and teams of researchers in academic, industrial, and military institutions, the branches of the history have become tangled and too numerous to describe exhaustively. Since the 1950s, it has become increasingly difficult to assign credit for computer breakthroughs to individual inventors.

Although individual contributors to the past two or three decades of computer research development have been abundant, the people who have been able to see some kind of overall direction to the fast, fragmented progress of recent years have been sparse. Just as the earliest logicians and mathematicians didn't know their thoughts would end up as part of a machine, the vast majority of the engineers and programmers of the 1960s were unaware that their machines had anything to do with human thought. The latter-day computer pioneers in the middle chapters of this book were among the few who played central roles in the development of personal computing. Like their predecessors, these people tried to create a kind of mental lever. Unlike most of their predecessors, they were also trying to design a tool that the entire population might use.

Where the original software patriarchs solved various problems in the creation of the first computers, the personal computer pioneers struggled with equally vexing problems involved in using computers to create leverage for human intellect, the way wheels and dynamos create leverage for human muscles. Where the patriarchs were out to *create* computation, the pioneers sought to *transform* it:

• **J.C.R. Licklider,** an experimental psychologist at MIT who became the director of the Information Processing Techniques Office of the U.S. Defense Department's Advanced Research Projects Agency (ARPA), was the one man whose vision enabled hundreds of other like-minded computer designers to pursue a whole new direction in hardware and software development. In the early 1960s, the researchers funded by Licklider's programs reconstructed computer science on a new and higher level, through an approach known as *time-sharing.*

Although their sponsorship was military, the people Licklider hired or supported were working toward a transformation that he and they believed to be social as well as technological. Licklider saw the new breed of interactive computers his project directors were creating as the first step toward an entirely new kind of human communication capability.

• **Doug Engelbart** started thinking about building a thought-amplifying device back when Harry Truman was President, and he has spent the past thirty years stubbornly pursuing his original vision of building a system for augmenting human intellect. At one point in the late 1960s, Engelbart and his crew of infonauts demonstrated to the assembled cream of computer scientists and engineers how the devices most people then used for performing calculations or keeping track of statistics could be used to enhance the most creative human activities.

His former students have gone on to form a disproportionate part of the upper echelons of today's personal-computer designers. Partially because of the myopia of his contemporaries, and partially because of his almost obsessive insistence on maintaining the purity of his original vision, most of Engelbart's innovations have yet to be adopted by the computer orthodoxy.

• **Robert Taylor,** at the age of thirty-three, became director of the ARPA office created by Licklider, thus launching his career in a new and much-needed field—the shaping of large-scale, long-term, human-computer research campaigns. He became a "people collector," looking for those computer researchers whose ideas might have been ignored by the orthodoxy, but whose projects promised to boost the state of computer systems by orders of magnitude.

Between 1965 and 1969, Taylor was responsible for initiating the next phase of interactive computer design—the construction of a communication medium to link the time-sharing communities all over the country into an unprecedented multicomputer, multimind entity known as the *ARPAnet*. In 1970, the majority of the top computer systems designers were once again collected by Taylor into one organization—Xerox Corporation's Palo Alto Research Center (PARC)—attracted by the promise of ten years of virtually unlimited funding to build the first true personal computers and communication network.

• **Alan Kay** was one of television's original quiz kids. He learned to read at the age of two and a half, barely managed to avoid being thrown out of school and the Air Force, and ended up as a graduate student at one of the most important centers of ARPA research. In the 1970s, Kay was one of the guiding software spirits of PARC's *Alto* project (the first true personal computer) and the chief architect of *Smalltalk*, a new kind of computer language. He started the 1980s as director of Atari Corporation's long-term research effort, and in 1984 he left Atari to become a "research fellow" for Apple Corporation.

Along with his hard-won credentials as one of the rare original thinkers who is able to implement his thoughts via the craft of software design, Kay also has a reputation as a lifelong insubordinate. Since the first time he was thrown out of a classroom for knowing more than the teacher, Kay's avowed goal has been to build a "fantasy amplifier" that anyone with an imagination could use to explore the world of knowledge on their own, a "dynamic medium for creative thought" that could be as useful and thought-provocative to children in kindergarten as it would be to scientists in a research laboratory.

Licklider, Engelbart, Taylor, and Kay are still at work, confident that many more of us will experience the same thrill that has kept them going all these years—what Licklider, still at MIT, calls the "religious conversion" to interactive computing. Engelbart works for Tymshare Corporation, marketing his "Augment" system to information workers. Taylor is setting up another computer systems research center, this time under the auspices of the Digital Equipment Corporation, and is collecting people once again, this time for a research effort that will bring computing into the twenty-first century. Kay, at Atari, continued to steer his team toward the fantasy amplifier, despite the fact that their mother company was often described in the news media as "seriously troubled." It is fair to assume that he will continue to work toward the same goal in his new association with Steve Jobs, chairman of Apple and a computer visionary of a more entrepreneurial bent.

The pioneers, although they are still at work, are not the final characters in the story of the computation quest. The next generations of innovators are already at work, and some of them are surprisingly young. Computer trailblazers in the past tended to make their marks early in life—a trend that seems to be continuing in the present. Kay, the former quiz kid, is now in his early forties. Taylor is in his early fifties, Engelbart in his late fifties, and Licklider in his sixties. Today, younger men and, increasingly, younger women have begun to take over the field professionally, while even younger generations are now living in their own versions of the future for fun, profit, and thrills.

The ones I call the "infonauts" are the older brothers and sisters of the adolescent hackers you read about in the papers. Most of them are in their twenties and thirties. They work for themselves or for some research institution or software house, and represent the first members of the McLuhan generation to use the technology invented by the von Neumann generation as tools to extend their imagination. From the science of designing what they call the "user interface"—where mind meets machine—to the art of building educational microworlds, the infonauts have been using their new medium to create the mass-media version we will use fifteen years from now.

• **Avron Barr** is a knowledge engineer who helps build the special computer programs known as *expert systems* that are apparently able to acquire knowledge from human experts and transfer it to other humans. These systems are now used experimentally to help physicians diagnose diseases, as well as commercially to help geologists locate mineral deposits and to aid chemists in identifying new compounds.

Although philosophers debate whether such programs truly "understand" what they are doing, and psychologists point out the huge gap between the narrowly defined kind of expertise involved in geology or diagnosis and the much more general "world knowledge" that all humans have, there is no denying that expert systems are valuable commodities. Avron Barr believes that they will evolve into more than expensive encyclopedias for specialists. In his midthirties and just starting his career in an infant technology, he dreams of creating an expert assistant in the art of helping people agree with one another.

• **Brenda Laurel,** also in her midthirties, is an artist whose medium exists at the boundary of Kay's and Barr's and Engelbart's specialties. Her goal is to design new methods of play, learning, and artistic expression into computer-based technologies. Like Barr, she believes that the applications of her research point toward more extensive social effects than just another success in the software market.

Brenda wants to use an expert system that knows what playwrights, composers, librarians, animators, artists, and dramatic critics know, to create a world of sights and sounds in which people can learn about flying a spaceship or surviving in the desert or being a blue whale by experiencing space-desert-whale simulated microworlds in person.

• **Ted Nelson** is a dropout, gadfly, and self-proclaimed genius who self-published *Computer Lib*, the best-selling underground manifesto of the microcomputer revolution. His dream of a new kind of publishing medium and continuously updated world-library threatens to become the world's longest software project. He's wild and woolly, imaginative and hyperactive, has problems holding jobs and getting along with colleagues, and was the secret inspiration to all those subteenage kids who lashed together homebrew computers or homemade programs a few years back and are now the ruling moguls of the microcomputer industry.

Time will tell whether he is a prophet too far ahead of his time, or just a persistent crackpot, but there is no doubt that he has contributed a rare touch of humor to the often too-serious world of computing. How can you not love somebody who says "they should have called it an oogabooga box instead of a computer"?

Despite their differences in background and personality, the computer patriarchs, software pioneers, and the newest breed of infonauts seem to share a distant focus on a future that they are certain the rest of us will see as clearly as they do—as soon as they turn what they see in their mind's eye into something we can hold in our hands. What did they see? What will happen when their visions materialize in our homes? And what do contemporary visionaries see in store for us next?

CHAPTER TWO

The First Programmer Was a Lady

Over a hundred years before a monstrous array of vacuum tubes surged into history in an overheated room in Pennsylvania, a properly attired Victorian gentleman demonstrated an elegant little mechanism of wood and brass in a London drawing room. One of the ladies attending this demonstration brought along the daughter of a friend. She was a teenager with long dark hair, a talent for mathematics, and a weakness for wagering on horse races. When she took a close look at the device and realized what this older gentleman was trying to do, the young lady surprised them all by joining him in an enterprise that might have altered history, had they succeeded.

Charles Babbage and his accomplice, Lady Lovelace, came very close to inventing the computer more than a century before American engineers produced ENIAC. The story of the "Analytical Engine" is a tale of two extraordinarily gifted and ill-fated British eccentrics whose biographies might have been fabrications of Babbage's friend Charles Dickens, if Dickens had been a science-fiction writer. Like many contemporary software characters, these computer pioneers of the Victorian age attracted as much attention with their unorthodox personal lives as they did with their inventions.

One of Babbage's biographies is entitled *Irascible Genius*. He was indeed a genius, to judge from what he planned to achieve as well as what he did achieve. His irascibility was notorious. Babbage was thoroughly British, stubbornly eccentric, tenaciously visionary, sometimes scatterbrained, and quite wealthy until he sank his fortune into his dream of building a calculating engine.

Babbage invented the cowcatcher—that metal device on the front of steam locomotives that sweeps errant cattle out of the way. He also devised a means of analyzing entire industries, a method for studying complex systems that became the foundation of the field of *operational research* a hundred years later. When he applied his new method of analysis to a study of the printing trade, his publishers were so offended that they refused to accept any more of his books.

Undaunted, he applied his new method to the analysis of the postal system of his day, and proved that the cost of accepting and assigning a value to every piece of mail according to the distance it had to travel was far more expensive than the cost of transporting it. The British Post Office boosted its capabilities instantly and economically by charging a flat rate, independent of the distance each piece had to travel—the "penny post" that persists around the world to this day.

Babbage devised the first speedometer for railroads, and he published the first comprehensive treatise on actuarial theory (thus helping to create the insurance industry). He invented and solved ciphers and made skeleton keys for "unpickable locks"—an interest in cryptanalysis that he shared with later computer builders. He was the first to propose that the weather of past years could be discovered by observing cycles of tree rings. And he was passionate about more than a few crackpot ideas that history has since proved to be nothing more than crackpot ideas.

His human relationships were as erratic as his intellectual adventures, to judge from the number of lifelong public feuds Babbage was known to have engaged in. Along with his running battles with the Royal Societies, Babbage carried on a long polemic against organ-grinders and street musicians. Babbage would write letters to editors about street noise, and half the organ-grinders in London took to serenading under Babbage's window when they were in their cups. One biographer, B. V. Bowden, noted that "It was the tragedy of the man that, although his imagination and vision were unbounded, his judgement by no means matched them, and his impatience made him intolerant of those who failed to sympathize with his projects." [1]

Babbage dabbled in half a dozen sciences and traveled with a portable laboratory. He was also a supreme nit-picker, sharp-eyed and cranky, known to write outraged letters to publishers of mathematical tables,

Charles Babbage, cranky eccentric genius, inventor of the ancestral digital computer—the "Analytical Engine." (Crown Copyright. Courtesy of the Charles Babbage Institute, University of Minnesota.)

upbraiding them for obscure inaccuracies he had uncovered in the pursuit of his own calculations. A mistake in a navigational table, after all, was a matter of life and death for a seafarer. And a mistake in a table of logarithms could seriously impede the work of a great mind such as his own.

His nit-picking indirectly led Babbage to invent the ancestor of today's computers. As a mathematician and astronomer of no small repute, he resented the time he had to spend poring over logarithm tables, culling all the errors he knew were being perpetuated upon him by "elderly Cornish clergymen, who lived on seven figure logarithms, did all their work by hand, and were only too apt to make mistakes." [2]

Babbage left a cranky memoir entitled *Passages from the Life of a Philosopher*—a work described by computer pioneer Herman Goldstine as "a set of papers ranging from the sublime to the ridiculous, from profundities to nonsense in plain bad taste. Indeed much of Babbage's career is of this sort. It is a wonder he had as many good and loyal friends when his behavior was so peculiar." [3]

In *Passages*, Babbage noted this about the original inspiration for his computing machines: [4]

> The earliest idea that I can trace in my own mind of calculating arithmetical tables by machinery rose in this manner:
> One evening I was sitting in the rooms of the Analytical Society at Cambridge, my head leaning forward on the table in a kind of

dreamy mood, with a Table of logarithms lying open before me. Another member, coming into the room, and seeing me half asleep, called out, "Well, Babbage, what are you dreaming about?" To which I replied, "I am thinking that all these Tables (pointing to the logarithms) might be calculated by machinery."

In 1822, Babbage triumphantly demonstrated to the Royal Astronomical Society a small working model of a machine, consisting of cogs and wheels and shafts. The device was capable of performing polynomial equations by calculating successive differences between sets of numbers. He was awarded the society's first gold medal for the paper that accompanied the presentation.

In that paper, Babbage described his plans for a much more ambitious "Difference Engine." In 1823, the British government awarded him the first of many grants that were to continue sporadically and controversially for years to come. Babbage hired a master machinist, set up shop on his estate, and began to learn at first hand how far ahead of his epoch's technological capabilities his dreams were running.

The Difference Engine commissioned by the British government was quite a bit larger and more complex than the model demonstrated before the Royal Astronomical Society. But the toolmaking art of the time was not yet up to the level of precision demanded by Babbage's design. Work continued for years, unsuccessfully. The triumphal demonstration at the beginning of his enterprise looked as if it had been the high point of Babbage's career, followed by a stubborn and prolonged decline. The British government finally gave up financing the scheme.

Babbage, never one to shy away from conflict with unbelievers over one of his cherished ideas, feuded over the Difference Engine with the government and with his contemporaries, many of whom began to make sport of mad old Charley Babbage. While he was struggling to prove them all wrong, he conceived an idea for an even more ambitious invention. Babbage, already ridiculously deep in one visionary development project, began to dream up another one. In 1833, he came up with something far more complex than the device he had failed to build in years of expensive effort.

If one could construct a machine for performing one kind of calculation, Babbage reasoned, would it be possible to construct a machine capable of performing *any* kind of calculation? Instead of building many small machines to perform many different kinds of calculation, would it be possible to make the parts of *one* large machine perform different tasks at different times, by changing the order in which the parts interact?

A model of a portion of the Analytical Engine, owned by the National Physical Laboratories, Teddington, England. (Crown Copyright. Courtesy of the Charles Babbage Institute, University of Minnesota.)

Babbage had stumbled upon the idea of a universal calculating machine, an idea that was to have momentous consequences when Alan Turing—another brilliant, eccentric, British mathematician who was tragically ahead of his time—considered it again in the 1930s. Babbage called his hypothetical master calculator the "Analytical Engine." The same internal parts were to be made to perform different calculations, through the use of different "patterns of action" to reconfigure the order in which the parts were to move for each calculation. A detailed plan was made, and redrawn, and redrawn once again.

The central unit was the "mill," a calculating engine capable of adding numbers to an accuracy of 50 decimal places, with a speed and reliability guaranteed to lay the Cornish clergymen calculators to rest. Up to one thousand different 50-digit numbers could be stored for later reference in the memory unit Babbage called the "store." To display the result, Babbage designed the first automated typesetter.

Numbers could be put into the store from the mill or from the punched-card input system Babbage adapted from French weaving machines. In addition, cards could be used to enter numbers into the mill and specify the calculations to be performed on the numbers as well. By using these cards properly, the mill could be instructed to temporarily place results in the store, then return the stored numbers to the mill for later procedures. The final component of the Analytical Engine was a card-reading device that was, in effect, a control and decision-making unit.

A working model was eventually built by Babbage's son. Babbage himself never lived to see the Analytical Engine. Toward the end of his life, a visitor found that Babbage had filled nearly all the rooms of his large house with abandoned models of his engine. As soon as it looked as if one means of constructing his device might actually work— Babbage thought of a new and better way of doing it.

The four subassemblies of the Analytical Engine functioned very much like analogous units in modern computing machinery. The mill was the analog of the central processing unit of a digital computer and the store was the memory device. Twentieth-century programmers would recognize the printer as a standard output device. It was the input device and control unit, however, that made it possible to move beyond calculation toward true computation.

The input portion of the Analytical Engine was an important milestone in the history of programming. Babbage borrowed the idea of punched-card programming from the French inventor Jacquard, who had triggered a revolution in the textile industry by inventing a mechanical method of weaving patterns in cloth. The weaving machines used arrays of metal rods to automatically pull threads into position. To create patterns, Jacquard's device interposed a stiff card, with holes punched in it, between the rods and the threads. The card was designed to block some of the rods from reaching the thread on each pass; the holes in the card allowed only certain rods from the complete array to carry threads into the loom. Each time the shuttle was thrown, a new card would appear in the path of the rods. Thus, once the directions for specific woven patterns were translated into patterns of holes punched into cards, and the cards were arranged in the proper order to present to the card-reading device, the cloth patterns could be preprogrammed and the entire weaving process could be automated.

These cards struck Babbage as the key to automated calculation. Here was a tangible means of controlling those frustratingly abstract "patterns of action": Babbage put the step-by-step instructions for complicated calculations into a coded series of holes punched into sets of cards that would change the way the mill worked at each step. Arrange the correctly coded cards in the right way, and you've replaced a platoon of elderly Cornish gentlemen. Change the cards, and you replace an entire army of them.

During his crusade to build the devices that he saw in his mind's eye but was somehow never able to materialize in wood and brass, Babbage met a woman who was to become his companion, colleague, conspirator, and defender. She saw immediately what Babbage intended to do with his Analytical Engine, and she helped him construct the software for it. Her work with Babbage and the essays she wrote about the possibilities of the engine established Augusta Ada Byron, Countess of Lovelace, as a patron saint if not a founding parent of the art and science of programming.

Ada's father was none other than Lord Byron, the most scandalous public character of his day. His separation from Ada's mother was one of the most widely reported domestic episodes of the era, and Ada never saw her father after she was one month old. Byron wrote poignant passages about Ada in some of his poetry, and she asked to be buried next to him—probably to spite her mother, who outlived her. Ada's mother, portrayed by biographers as a vain and overbearing Victorian figure, thought a daily dose of laudanum-laced "tonic" would be the perfect cure for her beautiful, outspoken daughter's nonconforming behavior, and thus forced an addiction on her!

Ada exhibited her mathematical talents early in life. One of her family's closest friends was Augustus De Morgan, the famous British logician. She was well tutored, but always seemed to thirst for more knowledge than her tutors could provide. Ada actively sought the perfect mentor, whom she thought she found in a contemporary of her mother's— Charles Babbage.

Mrs. De Morgan was present at the historic occasion when the young Ada Byron was first shown a working model of the Difference Engine, during a demonstration Babbage held for Lady Byron's friends. In her memoirs, Mrs. De Morgan remembered the effect the contraption had on Augusta Ada: "While the rest of the party gazed at this beautiful instrument with the same sort of expression and feeling that some savages are said to have shown on first seeing a looking glass or hearing a gun, Miss Byron, young as she was, understood its working and saw the great beauty of the invention." [5]

Such parlor demonstrations of mechanical devices were in vogue among the British upper classes during the Industrial Revolution. While her elders tittered and gossiped and failed to understand the difference between this calculator and the various water pumps they had observed at other demonstrations, young Ada began to knowledgeably poke and probe various parts of the mechanism, thus becoming the world's first computer whiz kid.

Ada was one of the few to recognize that the Difference Engine was altogether a different sort of device than the mechanical calculators of the past. Whereas previous devices were *analog* (performing calculation by means of *measurement*), Babbage's was *digital* (performing calculation by means of *counting*). More importantly, Babbage's design combined arithmetic and logical functions. (Babbage eventually discovered the new work on the "algebra of logic" by De Morgan's friend George Boole—but, by then, it was too late for Ada.)

Ada, who had been tutored by De Morgan, the foremost logician of his time, had ideas of her own about the possibilities of what one might do with such devices. Of Ada's gift for this new type of partially mathematical, partially logical exercise, Babbage himself noted: "She seems to understand it better than I do, and is far, far better at explaining it."

At the age of nineteen, Ada married Lord King, Baron of Lovelace. Her husband was also something of a mathematician, although his talents were far inferior to those of his wife. The young Countess Lovelace continued her mathematical and computational partnership with Babbage, resolutely supporting what she knew to be a solid idea, at a time when less-foresighted members of the British establishment dismissed Babbage as a crank.

Babbage toured the Continent in 1840, lecturing on the subject of the device he never succeeded in building. In Italy, a certain Count Menabrea in Italy took extensive notes at one of the lectures and published them in Paris. Ada translated the notes from French to English and composed an addendum which was more than twice as long as the text she had translated. When Babbage read the material, he urged Ada to publish her notes in their entirety.

Lady Lovelace's published notes are still understandable today and are particularly meaningful to programmers, who can see how truly far

Ada Byron, Lady Lovelace, Babbage's accomplice and the world's first computer programmer. (Crown Copyright. Courtesy of the Charles Babbage Institute, University of Minnesota.)

ahead of their contemporaries were the Analytical Engineers. Professor
B. H. Newman in the *Mathematical Gazette* has written that her observations "show her to have fully understood the principles of a programmed
computer a century before its time." [6]

Ada was especially intrigued by the mathematical implications of the
punched pasteboard cards that were to be used to feed data and equations
to Babbage's devices. Ada's essay, entitled "Observations on Mr. Babbage's Analytical Engine," includes more than one prophetic passage,
unheeded by most of her contemporaries, but which have grown in
significance with the passage of a century: [7]

> The distinctive characteristic of the Analytical Engine, and that
> which has rendered it possible to endow mechanism with such exten-
> sive faculties as bid fair to make this engine the executive right-hand
> of algebra, is the introduction into it of the principle which Jacquard
> devised for regulating, by means of punched cards, the most compli-
> cated patterns in the fabrication of brocaded stuffs. It is in this that
> the distinction between the two engines lies. Nothing of the sort
> exists in the Difference Engine. We may say most aptly that the
> Analytical Engine *weaves algebraical patterns* just as the Jacquard-
> loom weaves flowers and leaves. . . .
>
> The bounds of *arithmetic* were, however, outstepped the moment
> the idea of applying the cards had occurred; and the Analytical Engine
> does not occupy common ground with mere "calculating machines."
> It holds a position wholly its own; and the considerations it suggests
> are most interesting in their nature. In enabling mechanism to combine
> together *general* symbols, in successions of unlimited variety and ex-
> tent, a uniting link is established between the operations of matter
> and the abstract mental processes of the *most abstract* branch of
> mathematical science. A new, a vast, and a powerful language is devel-
> oped for the future use of analysis, in which to wield its truths so
> that these may become of more speedy and accurate practical applica-
> tion for the purposes of mankind than the means hitherto in our
> possession have rendered possible. Thus not only the mental and the
> material, but the theoretical and the practical in the mathematical
> world, are brought into more intimate and effective connexion with
> each other. We are not aware of its being on record that anything
> partaking of the nature of what is so well designated the *Analytical*
> Engine has been hitherto proposed, or even thought of, as a practical
> possibility, any more than the idea of a thinking or of a reasoning
> machine.

As a mathematician, Ada was excited by the possibility of automating
laborious calculations. But she was far more interested in the principles

underlying the programming of these devices. Had she not died so young, it is possible that Ada could have advanced the nineteenth-century state of the art to the threshold of true computation.

Even though the Engine was yet to be built, Ada experimented with writing sequences of instructions. She noted the value of several particular tricks in this new art, tricks that are still essential to modern computer languages—*subroutines, loops* and *jumps*. If your object is to weave a complex calculation out of subcalculations, some of which may be repeated many times, it is tedious to rewrite a sequence of a dozen or a hundred instructions over and over. Why not just store copies of often-used subcalculations, or subroutines, in a "library" of procedures for later use? Then your program can "call" for the subroutine from the library automatically, when your calculation requires it. Such libraries of subprocedures are now a part of virtually every high-level programming language.

Analytical Engines and digital computers are very good at doing things over and over many times, very quickly. By inventing an instruction that backs up the card-reading device to a specified previous card, so that the sequence of instructions can be executed a number of times, Ada created the loop—perhaps the most fundamental procedure in every contemporary programming language.

It was the conditional jump that brought Ada's gifts as a logician into play. She came up with yet another instruction for manipulating the card-reader, but instead of backing up and repeating a sequence of cards, this instruction enabled the card-reader to jump to another card in any part of the sequence *if* a specific condition was satisfied. The addition of that little "if" to the formerly purely arithmetic list of operations meant that the programs could do more than calculate. In a primitive but potentially meaningful way, the Engine could now make *decisions*.

She also noted that machines might someday be built with capabilities far beyond those possible with Victorian era technology, and speculated about the possibility of whether such machines could ever achieve intelligence. Her argument against artificial intelligence, set forth in her "Observations," was immortalized almost a century later by another software prophet, Alan Turing, who dubbed her line of argument "Lady Lovelace's Objection." It is an objection that is still frequently heard in debates about machine intelligence: "The Analytical Engine," Ada wrote, "has no pretensions whatever to originate anything. It can do whatever we know how to order it to perform." [8]

It is not known how and when Ada became involved in her clandestine and disastrous gambling ventures. No evidence has ever been produced that Babbage had anything to do with introducing Ada to what was to become her lifelong secret vice. For a time, Lord Lovelace shared Ada's obsession, but after incurring significant losses he stopped. She continued, clandestinely.

Babbage became deeply involved in Ada's gambling toward the end of her life. For her part, Ada helped Babbage in more than one scheme to raise money to construct the Analytical Engine. It was a curious mixture of vice, high intellectual adventure, and bizarre entrepreneurship. They built a tic-tac-toe machine, but gave up on it as a moneymaking venture when an adviser assured them that P. T. Barnum's General Tom Thumb had sewn up the market for traveling novelties. Ironically, although Babbage's game-playing machines were commercial failures, his theoretical approach created a foundation for the future science of game theory, scooping even that twentieth-century genius John von Neumann by about a hundred years.

It was Charley and Ada's attempt to develop an infallible system for betting on the ponies that brought Ada to the sorry pass of twice pawning her husband's family jewels, without his knowledge, to pay off blackmailing bookies. At one point, Ada and Babbage—never one to turn down a crazy scheme—used the existing small-scale working model of the Difference Engine to perform the calculations required by their complex handicapping scheme. The calculations were based on sound approaches to the theory of handicapping, but as the artificial intelligentsia were to learn over a century later, even the best modeling programs have trouble handling truly complex systems. They lost big. To make matters worse, when she compounded her losses Ada had to turn to her mother, who was not the most forgiving soul, to borrow the money to redeem the Lovelace jewels before her husband could learn of their absence.

Ada died of cancer at the age of thirty-six. Babbage outlived her by decades, but without Ada's advice, support, and sometimes stern guidance, he was never able to complete his long-dreamed-of Analytical Engine. Because the toolmaking art of his day was not up to the tolerances demanded by his designs, Babbage pioneered the use of diamond-tipped tools in precision-lathing. In order to systematize the production of components for his Engine, he devised methods to mass-manufacture interchangeable parts and wrote a classic treatise on what has since become known as "mass production."

Babbage wrote books of varying degrees of coherence, made breakthroughs in some sciences and failed in others, gave brilliant and renowned dinner parties with guests like Charles Darwin, and seems to have ended up totally embittered. Bowden noted that "Shortly before Babbage died he told a friend that he could not remember a single completely happy day in his life: 'He spoke as if he hated mankind in general, Englishmen in particular, and the English Government and Organ Grinders most of all.' " [9]

While Ada Lovelace has been unofficially known to the inner circles of programmers since the 1950s, when card-punched batch-processing was not altogether different from Ada's kind of programming, she has been relatively unknown outside those circles until recently. In the 1970s, the U.S. Department of Defense officially named its "superlanguage" after her.

Although it came too late to assist in the original design of the Analytical Engine, yet another discovery that was to later become essential to the construction of computers was made by a contemporary of Babbage and Lovelace. The creation of an algebra of symbolic logic was the work of another mathematical prodigy and British individualist, but one who worked and lived in a different world, far away from the parlors of upper-class London.

A seventeen-year-old Englishman by the name of George Boole was struck by an astonishing revelation while walking across a meadow one day in 1832. The idea came so suddenly, and made such a deep impact on his life, that it led Boole to make pioneering if obscure speculations about a heretofore unsuspected human faculty that he called "the unconscious." Boole's contribution to human knowledge was not to be in the field of psychology, however, but in a field of his own devising. As Bertrand Russell remarked seventy years later, Boole invented pure mathematics.

Although he had only recently begun to study mathematics, the teenage George Boole suddenly saw a way to capture some of the power of human reason in the form of an algebra. And Boole's equations actually worked when they were applied to logical problems. But there was a problem, and it wasn't in Boole's concept. The problem, at the time, was that nobody cared. Partly because he was from the wrong social class, and partly because most mathematicians of his time knew very little about logic, Boole's eventual articulation of this insight didn't cause much commotion when he published it. His revelation was largely ignored for generations after his death.

When the different parts of computer technology converged unexpectedly a hundred years later, electrical engineers needed mathematical tools to make sense of the complicated machinery they were inventing.

The networks of switches they created were electrical circuits whose behavior could be described and predicted by precise equations. Because patterns of electrical pulses were now used to encode logical operations like "and," "or," and the all-important "if," as well as the calculator's usual fare of "plus," "minus," "multiply," and "divide," there arose a need for equations to describe the logical properties of computer circuits.

Ideally, the same set of mathematical tools would work for both electrical and logical operations. The problem in the late 1930s was that nobody knew of any mathematical operations that had the power to describe both logical and electrical networks. Then the right kind of mind looked in the right place. An exceptionally astute graduate student at MIT named Claude Shannon, who later invented information theory, found Boole's algebra to be exactly what the engineers were looking for.

Without Boole, a poverty-stricken, self-taught mathematics teacher who was born the same year as Ada, the critical link between logic and mathematics might never have been accomplished. While the Analytical Engine was an inspiring attempt, it had remarkably little effect on the later thinkers who created modern computers. Without Boolean algebra, however, computer technology might never have progressed to the electronic speeds where truly interesting computation becomes possible.

Boole was right about the importance of his vision, although he wouldn't have known what to do with a vacuum tube or a switching circuit if he saw one. Unlike Babbage, Boole was not an engineer. What Boole discovered in that meadow and worked out on paper two decades later was destined to become the mathematical linchpin that coupled the logical abstractions of software with the physical operations of electronic machines.

Between them, Babbage's and Boole's inspirations can be said to characterize the two different kinds of motivation that caused imaginative minds over the centuries to try and eventually to succeed in building a computer. On the one side are scientists and engineers, who would always yearn for a device to take care of tedious computations for them, freeing their thoughts for the pursuit of more interesting questions. On the other side is the more abstract desire of the mathematical mind to capture the essence of human reason in a set of symbols.

Ada, who immediately understood Babbage's models when she saw them, and who was tutored by De Morgan, the one man in the world best equipped to understand Boole, was the first person to speculate at any length about the operations of machines capable of performing logical as well as numerical operations. Boole's work was not published until after Lady Lovelace died. Had Ada lived but a few years longer, her powerful intuitive grasp of the principles of programming would have been immeasurably enhanced by the use of Boolean algebra.

Babbage and Lovelace were British aristocrats during the height of the Empire. Despite the derision heaped on Babbage in some quarters for his often-peculiar public behavior, he counted the Duke of Wellington, Charles Dickens, and Prince Albert among his friends. Ada had access to the best tutors, the finest laboratory equipment, and the latest books. They were both granted the leisure to develop their ideas and the privilege of making fools of themselves in front of the Royal Society, if they desired.

Boole was the son of a petty shopkeeper, which wasn't the best route to a good scientific education. At the age of sixteen, his family's precarious financial situation obliged Boole to secure modest employment as a schoolteacher. Faced with the task of teaching his students something about mathematics, and by now thoroughly Lincolnesque in his self-educating skills, Boole set out to learn mathematics. He soon learned that it was the most cost-effective intellectual endeavor for a man of his means, requiring no laboratory equipment and a fairly small number of basic books. At seventeen he experienced the inspiration that was to result in his later work, but he had much to learn about both mathematics and logic before he was capable of presenting his discovery to the world.

At the age of twenty he discovered something that the greatest mathematicians of his time had missed—an algebraic theory of invariance that was to become an indispensable tool for Einstein when he formulated the theory of relativity. In 1849, after his long years as an elementary-school teacher, Boole's mathematical publications brought him an appointment as professor of mathematics at Queen's College, Cork, Ireland. Five years later, he published *An Investigation of the Laws of Thought, on Which Are Founded the Mathematical Theories of Logic and Probabilities.*

Formal logic had been around since the time of the Greeks, most widely known in the syllogistic form perfected by Aristotle, the simplified version of which most people learn no more than: "All men are mortal. Socrates is a man. Therefore Socrates is mortal." After thousands of years in the same form, Aristotelian logic seemed doomed to remain on the outer boundaries of the metaphysical, never to break through into the more concretely specified realm of the mathematical, because it was still just a matter of *words*. The next level of symbolic precision was missing.

For over a thousand years, the only logic-based system that was expressible in symbols rigorous and precise enough to be called "mathematical" had been the geometry set down by Euclid. Just as Euclid set down the basic statements and rules of geometry in axioms and theorems about spatial figures, Boole set down the basics of logic in algebraic symbols. This was no minor ambition. While knowledge of geometry is a widely useful tool for getting around in the world, Boole was convinced that logic was the key to human reason itself. He knew that he had found what every metaphysician from Aristotle to Descartes had overlooked. In his first chapter, Boole wrote: [10]

> 1. The design of the following treatise is to investigate the fundamental laws of those operations of the mind by which reasoning is performed; to give expression to them in the symbolical language of a Calculus, and upon this foundation to establish the science of Logic and construct its method . . . to collect from the various elements of truth brought to view in the course of these inquiries some probable intimations concerning the nature and constitution of the human mind. . . .
>
> 2. . . . To enable us to deduce correct inferences from given premises is not the only object of Logic . . . these studies have also an interest of another kind, derived from the light which they shed upon the intellectual powers. They instruct us concerning the mode in which language and number serve as instrumental aids to the processes of reasoning; they reveal to some degree the connexion between different powers of our common intellect; they set before us . . . the essential standards of truth and correctness—standards not derived from without, but deeply founded in the constitution of the human faculties. . . . To unfold the secret laws and relations of those high faculties of thought by which all beyond the merely perceptive knowledge of the world and of ourselves is attained or matured, is an object which does not stand in need of commendation to a rational mind.

Although his discovery had profound consequences in both pure mathematics and electrical engineering, the most important elements of Boole's algebra of logic were simple in principle. He used the algebra everybody learns in school as a starting point, made several small but significant exceptions to the standard rules of algebraic combination, and used his special version to precisely express the syllogisms of classical logic.

The concept Boole used to connect the two heretofore different thinking tools of logic and calculation was the idea of a mathematical system in which there were only two quantities, which he called "the Universe" and "Nothing" and denoted by the signs 1 and 0. Although he didn't know it at the time, Boole had devised a two-state system for quantifying logic that also happened to be a perfect method for analyzing the logic of two-state physical devices like electrical relays or vacuum tubes.

By using the symbols and operations specified, logical propositions could be reduced to equations, and the syllogistic conclusions could be computed according to ordinary algebraic rules. By applying purely mathematical operations, anyone who knew Boolean algebra could discover any conclusion that was logically contained in any set of specified premises.

Because syllogistic logic so closely resembles the thought processes of human reasoning, Boole was convinced that his algebra not only demonstrated a valid equivalence between mathematics and logic, but also represented a mathematical systematization of human thought. Since Boole's time, science has learned that the human instrument of reason is far more complicated, ambiguous, unpredictable, and powerful than the tools of formal logic. But mathematicians have found that Boole's mathematical logic is much more important to the foundation of their enterprise than they first suspected. And the inventors of the first computers learned that a simple system with only two values can weave very sophisticated computations indeed.

The construction of a theoretical bridge between mathematics and logic had been gloriously begun, but was far from completed by Boole's work. It remained for later minds to discover that although it is probably not true that the human mind resembles a machine, there is still great power to be gained by thinking about machines that resemble the operations of the mind.

Nineteenth-century technology simply wasn't precise enough, fast enough, or powerful enough for ideas like those of Babbage, Lovelace, and Boole to become practicalities. The basic science and the industrial capabilities needed for making several of the most important components of modern computers simply didn't exist. There were still important problems that would have to be solved by inventors rather than theorists.

The next important development in the prehistory of computation, and the last important contribution of the nineteenth century, had nothing to do with calculating tables of logarithms or devising laws of thought. The next thinker to advance the state of the art was Herman Hollerith, a nineteen-year-old employee of the United States Census Office. His role would have no effect on the important theoretical foundations of computing. Ultimately, his invention became obsolete. But his small innovation eventually grew into the industry that later came to dominate the commercial use of computer technology.

Hollerith made the first important American contribution to the evolution of computation when his superior at the Census Office set him to work on a scheme for automating the collection and tabulation of data. On his superior's suggestion, he worked out a system that used cards with holes punched in them to feed information to an electrical counting system.

The 1890 census was the point in history where the processing of *data* as well as the calculation of mathematical equations became the object of automation. As it turned out, Hollerith was neither a mathematician nor a logician, but a data processor. He was grappling, not with numerical calculation, but with the complexities of collecting, sorting, storing, and retrieving a large number of small items in a collection of information. Hollerith and his colleagues were unwitting forerunners of twentieth-century information workers, because their task had to do with finding a mechanical method to keep track of what their organization knew.

Hollerith was introduced to the task by his superior, John Shaw Billings, who had been worrying about the rising tide of information since 1870, when he was hired by the Census Office to develop new ways to handle large amounts of information. Since he was in charge of the collection and tabulation of data for the 1880 and 1890 census, Billings was acutely aware that the growing population of the nation was straining the ability of the government to conduct the constitutionally mandated survey every ten years. In the foreseeable future, the flood of information to be counted and sorted would take fifteen or twenty years to tabulate!

Like the stories about the origins of other components of computers, there is some controversy about the exact accreditation for the invention of the punched-card system. One account by a man named Willcox, who worked with both Billings and Hollerith in the Census Office, stated: [11]

> While the returns of the Tenth (1880) Census were being tabulated at Washington, Billings was walking with a companion through the office in which hundreds of clerks were engaged in laboriously transferring items of information from the schedules to the record sheets by the slow and heartbreaking method of hand tallying. As they were watching the clerks he said to his companion, 'There ought to be some mechanical way of doing this job, something on the principle of the Jacquard loom, whereby holes in a card regulate the pattern to be woven.' The seed fell on good ground. His companion was a

talented young engineer in the office who first convinced himself that the idea was practicable and then that Billings had no desire to claim or use it.

The "talented young engineer," of course, was Hollerith, who wrote this version in 1919: [12]

One Sunday evening at Dr. Billings' tea table, he said to me there ought to be a machine for doing the purely mechanical work of tabulating population and similar statistics. We talked the matter over and I remember . . . he thought of using cards with the description of the individual shown by notches punched in the edge of the card. . . . After studying the problem I went back to Dr. Billings and said that I thought I could work out a solution for the problem and asked him if he would go in with me. The Doctor said he was not interested any further than to see some solution of the problem worked out.

The system Hollerith put together used holes punched in designated locations on cardboard cards to represent the demographic characteristics of each person interviewed. Like Jacquard's and Babbage's cards, and the "player pianos" then in vogue, the holes in Hollerith's cards were meant to allow the passage of mechanical components. Hollerith used an electromechanical counter in which copper brushes closed certain electrical circuits if a hole was encountered, and did not close a circuit if a hole was not present.

An electrically activated mechanism increased the running count in each category by one unit every time the circuit for that category was closed. By adding sorting devices that distributed cards into various bins, according to the patterns of holes and the kind of tabulation desired, Hollerith not only increased the ability to keep up with large amounts of data, but created the ability to ask new and more complicated questions about the data. The new system was in place in time for the 1890 census.

Hollerith obtained a patent on the system that he had invented just in time to save the nation from drowning in its own statistics. In 1882–83, he was an instructor in mechanical engineering at the Massachusetts Institute of Technology, establishing the earliest link between that institution and the development of computer science and technology. In 1896, Hollerith set up the "Tabulating Machine Company" to manufacture both the cards and the card-reading machines. In 1900, Hollerith rented his equipment to the Census Bureau for the Twelfth Census.

Some years later, Hollerith's Tabulating Machine Company had become an institution known as "International Business Machines," run by a fellow named Thomas Watson, Senior. But there were two World Wars ahead, and several more thinkers—the most extraordinary of them all—still to come before a manufacturer of tabulating machines and punch cards would have anything to do with true computers. The modern-day concerns of this company—selling machines to keep track of the information that goes along with doing business—would have to wait for some deadly serious business to be transacted.

The War Department, not the Census Office or a business machine company, was the mother of the digital computer, and the midwives were many—from Alan Turing's British team who needed a special kind of computing device to crack the German code, to John von Neumann's mathematicians at Los Alamos who were faced with almost insurmountable calculations involved in building the atomic bomb, to Norbert Wiener's researchers who were inventing better and faster ways to aim antiaircraft fire, to the project at the Army Ballistic Research Laboratory that produced the Electronic Numerical Integrator and Calculator (ENIAC).

It would be foolish to speculate about what computers might become in the near future without realizing where they originated in the recent past. The historical record is clear and indisputable on this point: ballistics begat cybernetics. ENIAC, the first electronic digital computer, was originally built in order to calculate ballistic firing tables. When ENIAC's inventors later designed the first miniature computer, it was the BINAC, a device small enough to fit in the nose cone of an ICBM and smart enough to navigate by the position of the stars.

Although the the first electronic digital computer was constructed in order to produce more accurate weapons, the technology would not have been possible without at least one important theoretical breakthrough that had nothing to do with ballistics or bombs. The theoretical origins of computation are to be found, not in the search for more efficient weaponry, but in the quest for more powerful and elegant symbol systems.

The first modern computer was not a machine. It wasn't even a blueprint. The digital computer was conceived as a symbol system—the first *automatic* symbol system—not as a tool or a weapon. And the person who invented it was not concerned with ballistics or calculation, but with the nature of thought and the nature of machines.

CHAPTER THREE

The First Hacker and His Imaginary Machine

Throughout the winter of 1936, a young Cambridge don put the finishing touches on a highly technical paper about mathematical logic that he didn't expect more than a dozen people around the world to understand. It was an unusual presentation, not entirely orthodox by the rather rigid standards of his colleagues. The young man wasn't entirely orthodox, himself. Although his speech revealed his upper-middle class origins, his manner of dress, his erratic grooming, and his grating voice put off most of his peers. An outsider to the loftier academic-social circles of the university, he had few friends, preferring to spend his time at mathematics, chemistry experiments, chess puzzles, and long runs in the countryside.

Computation, when it was finally invented, a century after Babbage, did not come in the form of some new gadget in an inventor's workshop or a scientist's laboratory. The very possibility of building digital computers was given to the world in the form of an esoteric paper in a mathematics journal in 1936.[1] Nobody realized at the time that this peculiar discovery in the obscure field of metamathematics would eventually lead to a world-changing technology, although the young author, Alan Mathison Turing, knew he was on the track of machines that could simulate human thought processes.

That mathematics paper was a pivotal point in the cultural history of Western civilization. The first move in the intellectual game that resulted in digital computers was also one of the last moves in another game that had gone on for millennia. In Egypt and Babylonia, where the systems for measuring land and forecasting the course of the stars originated, only the priests and their chosen craftsmen were privileged to know the esoteric arts of reckoning. During the flowering of Greek civilization in the fifth and sixth centuries B.C., these protosciences were shaped into the mental tools known as *axiomatic* systems.

In an axiomatic system you start with premises that are known to be true, and rules that are known to be valid, in order to produce new statements that are guaranteed to be true. Conclusions can be reached by manipulating symbols according to sets of rules. Euclidean geometry is the classic example of the kind of generally useful tools made possible by formal axiomatic systems.

An axiomatic system is a tool for augmenting human thought. Except for rare "lightning" calculators, people are not able to add two six-figure numbers in their head. Give virtually all people over the age of ten a piece of paper and a pencil, however, and they'll tell you the answer in less than a minute. The magic ingredient that makes a schoolchild into a calculating machine is the kind of step-by-step recipe for performing a calculation that is known as an *algorithm*. The reason we know such algorithms work is because they are based on the formal system known as arithmetic, which we know to be true.

What Turing's paper did, and what made digital computers possible, resulted from the millennia-long effort to reduce the various formal systems to one basic system that underlies them all. Science—our civilization's preeminent system for gathering and validating knowledge—was built on mathematics, which was in turn a logical formalization of the primitive number theories of the Babylonians and Greeks. Computation was the unexpected result of the attempt to prove that mathematical truths could be reduced to logical truths.

At the same time that our civilization's methods for predicting and understanding the universe grew powerful as the result of these intellectual systems (i.e., science, mathematics, and logic), a few people continued to ask whether these same systems could be reduced to their basic components. If all sciences, when they become advanced enough, can be reduced to mathematical equations, is it possible to reduce mathematics to the most fundamental level of logic?

Since our certainty in the completeness and consistency of our knowledge system could depend on whether such a reduction was possible, it was very disconcerting to Western thinkers when evidence began to appear that there were exceptions, anomalies, paradoxes—holes in the structure of mathematics that might prevent any such grand reduction of formal systems. Those two intellectual quests—the effort to reduce mathematics to a fundamental, formal symbol system, and the attempt to patch up the paradoxes that cropped up during the pursuit of that grand reduction—led directly but unexpectedly to computation.

In the first decades of the twentieth century, mathematicians and logicians were trying to *formalize* mathematics. David Hilbert and John von Neumann set down the rules of formalism in the 1920s (as we shall see in the next chapter). Before Hilbert and von Neumann, Alfred North Whitehead and Bertrand Russell demonstrated in their *Principia Mathematica* that some aspects of human reasoning could be formally described, thus linking this awakened interest in mathematical logic to the ideas of the long-forgotten originator of the field, George Boole. The idea of *formal systems* was of particular interest, because it appeared to bridge the abstractions of mathematics and the mysteries of human thought.

A formal system is a rigidly defined kind of game that specifies rules for manipulating tokens. The qualifications for making a formal system are very much like the rules for any other game. To tell someone how to play a game, and for a set of rules to qualify as a formal system, the same three aspects of the game must be communicated—the nature of the tokens, a description of the starting position (or the starting layout of the "board"), and a listing of what moves are allowed in any given position. Chess, checkers, mathematics, and logic are examples of formal systems that satisfy these criteria. By the 1930s, the effort to reduce mathematics to logically secure foundations brought about several attempts to treat arithmetic—the branch of mathematics concerned with operations on numbers—as a formal system.

In 1936, at the age of twenty-four, Alan M. Turing established himself as one of the greatest mathematical prodigies of all time when he pointed out to his colleagues that it was possible to perform computations in number theory by means of a machine—a machine that embodied the rules of a formal system. Although the machine itself didn't exist as a working model, Turing emphasized from the beginning that such machines could actually be built. His finding was a milestone in the effort to formalize mathematics and, at the same time, a watershed in the history of the theory of computation.

In his brilliant solution to one of the key metamathematical problems posed by the formalists, Alan Turing described in precise mathematical terms how an *automatic* formal system with extremely simple rules of operation could have very powerful capabilities. An automatic formal system is a physical device which automatically manipulates the tokens of a formal system according to that system's rules. Turing's theoretical machine was both an example of his theory of computation and a proof that a certain kind of computing machine could, in fact, be constructed.

When he brought mathematics and logic together in the form of a machine, Turing made *symbol-processing* systems possible. He proposed that the vast majority of intellectual problems could be converted to the form "find a number n, such that . . . " Even more important than this provocative statement connecting the abstractions of intellect with the more concrete realm of numbers—an implication that still inspires the efforts of artificial intelligence researchers—was Turing's recognition that the numbers were more important as *symbols* in this case than as elements of mathematical calculations.

One of Turing's greatest insights was his understanding, from the very beginning, of something that the majority of the computer priesthood has yet to understand—the fact that *numbers are only one possible way of interpreting the internal states of an automatic formal system*. Babbage's "patterns of action" were now formalized with mathematical rigor. Turing's "states" provided the crucial metaphor for bridging the power of human cognition and the capabilities of machines.

What, Turing asked, does a human symbol processor do when performing a calculation? He decided that mental calculations consist of operations for transforming the input numbers into a series of intermediate *states* which progress from one to the next according to a fixed set of rules, until an answer is found. Sometimes, people use pencil and paper to keep track of the states of their calculations. The rules of mathematics require more rigid definitions than those provided by the fuzzily described human states of mind discussed by metaphysicians, so Turing concentrated on defining these states in a way that was so clear and unambiguous that the description could be used to command the operations of a machine.

Turing started with a precise description of a formal system, in the form of "instruction tables" describing which moves to make for every possible configuration of states of the system. He then proved that the description of these instructions, the steps of formal axiomatic systems like logic, and the machine states that make up the "moves" in an automatic formal system are all equivalent to one another. Such matters as formal systems and Turing machines sound very far away from what computers actually do, but in fact they underlie the entire technology of digital computers—which wasn't to come into existence until over a decade after Alan Turing published his epochal paper.

The process of computation was graphically depicted in Turing's paper when he asked the reader to consider a device that can read and write simple symbols on a paper tape that is divided into squares. The "reading/ writing head" can move in either direction along the tape, one square at a time, and a control unit that directs the actions of the head can interpret simple instructions about reading and writing symbols in squares. The single square that is "scanned" or "read" at each stage is known as the *active square*. Imagine that new sections of tape always can be added at either end of the existing tape, so it is potentially infinite.

Suppose the symbols are "X" and "O." Suppose that the device can erase either symbol when it reads it in the active square and replace it with the other symbol (i.e., erase an X and replace it with an O, and vice versa). The device also has the ability to move left or right, one square at a time, according to instructions interpreted by the control unit. The instructions cause a symbol to be erased, written, or left the same, depending on which symbol is read.

Any number of games can be constructed using these rules, but they would not all necessarily be meaningful. One of the first things Turing demonstrated was that some of the games constructed under these rules can be very sophisticated, considering how crude and automaton-like the primitive operations seem to be. The following example illustrates how this game can be used to perform a simple calculation.[2]

The rules of the game to be played by this Turing machine are simple: Given a starting position in the form of a section of tape with some Xs and Os on it, and a starting square indicated, the device is to perform the actions dictated by a list of instructions that are interpreted by its control unit. The device starts with the first instruction and follows the succeeding instructions one at a time until it reaches an instruction that forces it to stop. (If there is no explicit instruction in the table of instructions for a particular tape configuration, there is nothing that the machine can do when it reaches that configuration, so it has to stop.)

Each instruction specifies a particular action to be performed if there is a certain symbol on the active square at the time it is read. There are four different actions; they are the only legal moves of this game. They are

Replace O with X.
Replace X with O.
Go one square to the right.
Go one square to the left.

An example of an instruction is: "If there is an X on the active square, replace it with O." This instruction causes the machine to perform the second action listed above. In order to create a "game," we need to make a list that specifies the number of the instruction that is being followed at every step as well as the number of the instruction that is to be followed next. This is like saying: "The machine is now following (for example) instruction seven, and the instruction to be followed next is (for example) instruction eight."

We can write all four parts of an instruction (the number of the instruction being followed, which one of the two symbols is on the active square, which action to take, and the number of the instruction to follow next) in a coded form such as: "7XL8." This means: "This is instruction number seven, which specifies that if the active square contains an X, the machine will move one square to the left, and then proceed to instruction number eight."

Here is a series of instructions, given in both the coded form and the more English-like translation. Taken together, these instructions constitute an "instruction table" or a "program" that tells a Turing machine how to play a certain kind of game:

1XO2   (Instruction #1: if an X is on the active square, replace it with O, then execute instruction #2.)

2OR3   (Instruction #2: if an O is on the active square, go right one square and then execute instruction #3.)

3XR3   (Instruction #3: if an X is on the active square, go right one square and then execute instruction #3;

3OR4   but if an O is on the active square, go right one square and then execute instruction #4.)

4XR4   (Instruction #4: if an X is on the active square, go right one square and then execute instruction #4;

4OX5   but if an O is on the active square, replace it with X and then execute instruction #5.)

5XR5   (Instruction #5: if an X is on the active square, go right one square and then execute instruction #5;

5OX6	but if an O is on the active square, replace it with X and then execute instruction #6.)
6XL6	(Instruction #6: if an X is on the active square, go left one square and then execute instruction #6;
6OL7	but if an O is on the active square, go left one square and then execute instruction #7.)
7XL8	(Instruction #7: if an X is on the active square, go left one square and then execute instruction #8.)
8XL8	(Instruction #8: if an X is on the active square, go left one square and then execute instruction #8;
8OR1	but if an O is on the active square, go right one square and then execute instruction #1.)

Note that if there is an O on the active square in instruction #1 or #7, or if there is an X on the active square in instruction #2, the machine will stop.

In order to play the game (run the program) specified by the list of instructions, one more thing must be provided: a starting tape configuration. For our example, let us consider a tape with two Xs on it, bounded on both sides by an infinite string of Os. The changing states of the single tape are depicted here as a series of tape segments, one above the other. The active square for each state is denoted by a capital X or O. When the machine is started it will try to execute the first available instruction, instruction #1. The following series of actions will then occur:

Instruction	Tape	What the Machine Does
#1	...ooXxooooooo...	One (of two) Xs is erased.
#2	...ooOxooooooo...	
#3	...oooXooooooo...	Tape is scanned to the
#3	...oooxOoooooo...	right.
#4	...oooxoOooooo...	
#5	...oooxoXooooo...	Two Xs are written.
#5	...oooxoxOoooo...	
#6	...oooxoxXoooo...	
#6	...oooxoXxoooo...	Scanner returns to the
#6	...oooxOxxoooo...	other original X.
#7	...oooXoxxoooo...	
#8	...ooOxoxxoooo...	
#1	...oooXoxxoooo...	

#2	...oooOoxxoooo...	This X is erased.
#3	...ooooOxxoooo...	Scanner moves to the right
#4	...oooooXxoooo...	of the two Xs that were
#4	...oooooxXoooo...	written earlier.
#4	...ooooooxxOooo...	
#5	...ooooooxxXooo...	Two more Xs are written.
#5	...ooooooxxxOoo...	
#6	...ooooooxxxXoo...	
#6	...ooooooxxXxoo...	Scanner looks for any more
#6	...oooooxXxxoo...	original Xs.
#6	...oooooXxxxoo...	
#6	...ooooOxxxxoo...	
#7	...oooOoxxxxoo...	The machine stops because there is no instruction for #7 if O is being scanned.

This game may seem rather mechanical. The fact that it is mechanical was one of the points Turing was trying to make. If you look at the starting position, note that there are two adjacent Xs. Then look at the final position and note that there are four Xs. If you were to use the same instructions, but start with a tape that had five Xs, you would end up with ten Xs. This list of instructions is the specification for a calculating procedure that will double the input and display the output. It can, in fact, be done by a machine.

In essence, every Turing machine moves marks from one position on a tape to another position on a tape, in the way the procedure outlined above moved Xs and Os from square to square. These days, the marks can be electronic impulses in microcircuits, and the tape can be an array of memory locations in a memory chip, but the essential idea is the same. Turing proved that his hypothetical machine is an automated version of a formal system specified by the starting position (the pattern of Os and Xs on the tape at the beginning of the computation) and the rules (the instructions given by the instruction tables). The moves of the game are the changing states of the machine that correspond to the specified steps of the computation.

Turing then proved that *for any formal system, there exists a Turing machine that can be programmed to imitate it.* This kind of general formal system with the ability to imitate any other formal system was what Turing was getting at. These systems are now known as "universal Turing machines." The theory was first stated in a paper with the forbidding title "On Computable Numbers, with an Application to the Entscheidungsproblem." [3]

The Turing machine was a hypothetical device Turing invented on the way to settling a critical question about the foundations of mathematics as a formalized means of thinking. He showed that his device could solve infinitely many problems, but that there are some problems that cannot be solved because there is no way of predicting in advance whether or when the machine is going to stop. Here is where the parting of the ways between metamathematics and computation occurred.

Our simple example of a doubling program took only twenty-six steps. But there is no way of knowing whether or not other programs (which can be direct translations of theorems in number theory) will ever stop. By proving this, Turing made an equivalent point about all mechanical systems (i.e., systems in which the procedures are definite enough to be carried out by a machine).

Turing and his colleagues ended the long search for a logically certain basis underlying formal systems by making the shocking discovery that there are a number of important features of formal systems about which we can never be absolutely certain. Formal systems, by their very nature, have certain inherent limitations. At this point, the theory of computation became something more than an important branch of metamathematics, as the properties of formal systems faded into the background and the properties of machines emerged into history in a wholly unexpected and dramatic manner—because at the same time that Turing put a limit on the capabilities of formal systems, he showed that there is indeed such a thing as a universal formal system. And that is what a computer is, in the most basic sense.

The way the universal Turing machine imitates other Turing machines is as automatic as the way our doubling machine multiplies the input by two. Assuming that the control unit of the device is capable of interpreting simple instructions—something that had been a matter for toolmakers, not mathematicians, since Babbage's time—it is possible to encode a more complex list of instructions describing various Turing machines and put them onto the input tape, along with the starting position.

Just as the instructions followed by the machine can be stated in English (or German or French, etc.), or in an abbreviated form like "7XL8," they can be encoded in an even more primitive form. A code can be devised, using the same Xs and Os, that can uniquely represent every intruction and instruction table (program). Both the instructions and the data can be put onto the same tape. A universal Turing machine can then scan that coded tape and perform the function specified in the code (doubling the number on the data portion of the tape, in our example).

This code can be interpreted by a machine, a machine that *automatically* manipulates the tokens, given a list of instructions and a starting configuration. When the machine stops, you read the tape and you get the output of the program. In this case, you put the number you want to double into the starting configuration, then let the machine metaphorically clank away one square at a time, erasing and writing Os or Xs. When the machine stops, you count the Xs in the final tape configuration.

The list of instructions is what turns the universal Turing machine into the doubling machine. Mechanically, there is no difference between the two machines. The particular instructions described by the code are what the universal Turing machine operates upon. If you can describe, in similarly codable instructions, a machine for tripling, or extracting square roots, or performing differential equations, then your basic, dumb old universal Turing machine can *imitate* your tripling machine or square root machine.

That ability to imitate other machines is what led to computers. The numbers (or Xs and Os) on the tape aren't that important. They are only symbols for states of a process—markers in a "doubling game." The list of instructions (the program) is what enables the machine to double the input number. The instructions, not the symbols that keep track of the way they are carried out—the rules, not the markers—are what make the Turing machine work. Universal Turing machines are primarily *symbol manipulators*. And digital computers are universal Turing machines.

It isn't easy to think of the rules of a game as a kind of machine. The task is somewhat easier if you think about "mechanical processes" that are so clearly and specifically defined that a machine can perform them by referring to an instruction table. All universal Turing machines are functionally identical devices for following the program specified by an instruction table. The instruction tables can differ, and they can turn the universal Turing machine into many different kinds of machine. For this reason, the programs are sometimes called "virtual machines."

The distinction between a universal Turing machine and the many different Turing machines it is able to imitate is a direct analogy to digital computers. Like universal Turing machines, all digital computers are functionally identical. At the most basic level, every digital computer operates in the way our doubling machine did with the squares and 0s and Xs. Instead of building a different physical machine to solve different problems, it is more practical to describe to an instruction-following machine different *virtual* machines (programs) that use this one-square-at-a-time mechanical instruction-following process to solve complicated problems through a pattern of simple operations.

Following instructions is the nature of digital computers. The difference between a computer calculator and a computer typewriter, for example, lies in the instructions it follows—the coded description it is given of the virtual machine it is meant to imitate in order to perform a task. Since computers understand "bits" that can correspond to symbols like O and X, or 0 and 1, or "on" and "off," you can use these symbols to write descriptions that turn the general machine into the specific machine you want. That's what programmers do. They think of machines people might want to use, and figure out ways to describe those special machines to general machines—computers, that is.

It would be too time-consuming to achieve anything significant in programming if programmers had to spend their time thinking of ways to describe machines in strings of Os and Xs. The O and X code is similar to what is now called *machine language*, and a relatively small number of programmers are actually able to write programs in it. But what if you could build a virtual machine on top of a virtual machine? What if there were a coded program written in terms of Os and Xs, much like the system we described for the doubling machine, except this new system's task is to translate symbols that humans find easier to use and understand—instructions like "go left" or even "double this number"—into machine language?

Assembly language, a close relative of machine language except that it uses recognizable words instead of strings of Xs and Os, is a lot more manageable than machine language, so that's what most programmers use when they write video games or word processors. Assembly language makes it easier to manipulate the information in the "squares"—the memory cells of the computer—by using words instead of numbers. You use the translation program described above, called an *assembler*, to translate assembly language into machine language.

Every different microprocessor (the actual silicon chip hardware at the core of every modern computer) has a list of around a hundred primitive machine language operations—known as "firmware"—wired into it. When the assembler follows the instructions in the assembly language programs, using machine language to talk to the microprocessor, the virtual machine meets the actual machine, and the computer is able to accomplish the specified task for the human who started the whole process.

Since you have to accomplish tasks in assembly language by telling the computer very specifically where to find the information you want, when to move it into an "active square" called an *accumulator*, and where to store it when it is processed, writing anything complicated in assembly language can be a chore—like writing a book with semaphore flags, or measuring a city with a yardstick.

For example, to add two numbers in assembly language, you have to specify what the first number is and assign it to the accumulator, then you have to specify the second number and instruct the machine to add it to the number already in the accumulator. Then you have to specify where to store the answer, and issue step-by-step instructions on how to send the answer to your printer or monitor.

Obviously, it is easier to do the whole thing in a procedure like the one in BASIC: You simply type something on the computer keyboard, like "PRINT 2 + 3," and some part of the software takes care of accumulators and memory addresses. Your printer prints out "5," or it is displayed on your monitor, and the computer doesn't bother you with details about its internal operations.

At the core of every computer language is something very much like the doubling machine. Since it is possible to describe machines that describe machines, under the rules of the universal Turing machine game, it is possible to write a machine language program that describes a machine that can translate assembly language into machine language. Having done that, this new tool can be used to create yet another level of communication that is even more manageable than assembly language, by making a code-language that is still closer to English.

That last virtual machine—the English-like one—is called a high-level programmming language. *High-level* doesn't mean that a language is intellectually lofty, only that it is a virtual machine interpreted by a lower-level machine, which in turn may be interpreted by an even lower-level machine, until you get to the lowest level of on and off impulses that translate the Os and Xs into electronically readable form. BASIC and FORTRAN and the other languages that programmers work with are actually virtual machines that are described to the computer by other virtual machines equivalent in function to the assemblers mentioned above, known as *interpreters* and *compilers*.

The first compiler, however, was not to be written until 1953, seventeen years after Turing's theoretical paper was published in 1936. The emergence of the digital computer, based on the principles of Turing's machine, was stimulated by World War II, which was still four years in the future. In 1936, Claude Shannon had yet to discover that the algebra invented by George Boole to formalize logical operations was identical with the mathematics used to describe switching circuits. John von Neumann and his colleagues had yet to devise the concept of stored programming. Norbert Wiener hadn't formalized the description of feedback circuits in control systems. Several crucial electronic developments were yet to come.

Although only a half-dozen metamathematicians thought about such things during the 1930s, the notion of machines whose functions depend on the descriptions of how they operate happened to have one real-world application that suddenly became very important toward the end of the decade. In 1940, the British government developed an intense interest in Turing's theories.

A top-secret project code-named "Ultra," under the direction of an intelligence officer code-named "Intrepid," had captured and brought to London the secret German cipher machine known as "Enigma." The machine enabled the Nazi high command to send orders to field commanders in the form of an uncrackable code. Even though they had the machine in their hands, British intelligence was still baffled by the encoding mechanism. Even the best of the old-style cryptographers couldn't suggest a solution.

The British high command recruited brilliant mathematicians, engineers, and logicians, inadvertently creating one of the seminal research groups in the field that was to become known as artificial intelligence. Among them was Donald Michie, then only twenty-two, who was later to become the leading British machine intelligence researcher. Another very young colleague who later distinguished himself was I. J. Good, a prankster who once wrote to Her Majesty the Queen suggesting that he be made a peer of the realm, because then his friends would be forced to remark, "Good Lord, here comes Lord Good," when they saw him coming.

The place known as Bletchley Park is far less famous than Omaha Beach, but many historians contend that the European war was won, in large part, in a closely guarded Victorian mansion in Hertfordshire, England, by the group of thinkers who succeeded in breaking the German code. The brilliant, young, unorthodox code-crackers were housed near Bletchley Park while they performed their role in the top-secret operation. One of the code-breakers was twenty-eight-year-old Alan Turing.

Turing was eccentric, fun-loving, disheveled, painfully honest, erratic,

Alan Turing at the finish of a long-distance race. He often ran long distances with an alarm clock tied around his waist, as a timing device. (Courtesy of John Edward Leigh.)

introspective, prodigiously and elegantly brilliant, and somewhat inept socially. Turing was an early model of the similarly maladroit and analogously otherworldly computer hackers who were to come later: He was a sloppy dresser and a passionate chessplayer, fond of children's radio programs and dedicated to long-distance running. (Sometimes he timed himself with an alarm clock tied around his waist.) Even one of his few intimate friends described his speech as "a shrill stammer and crowing laugh which told upon the nerves even of his friends." [4]

He never quite got the hang of automobiles, which was probably safer, considering the way Turing's mind wandered far away from the realities of the roadway. He preferred the battered bicycle of the Cambridge don. The bicycle and his habit of running twenty or thirty miles to attend a meeting were the objects of sundry anecdotes about "the Prof," as Turing was known around Bletchley. He was once detained by the local constable for bicycling around in a gas mask, which Turing claimed alleviated his hay fever.

Turing and his colleagues at Bletchley Park ended up solving the Enigma enigma by devising a series of machines known as "bombes," "the Robinsons," and a culminating contraption known as "Colossus." Their purpose? To imitate "Enigma," of course!

The Bletchley Park devices were by no means universal machines by Turing's 1936 definition, but they did use important aspects of Turing's ideas. Using high-speed devices for feeding instructions encoded on paper tapes, and electrical circuitry for performing simple but tedious logical operations upon coded messages, the decoding machines started operating in 1943. The machines enabled the British to successfully crack Enigma's code, in part by imitating crucial functions of the enemy encoding machine.

The fact that these young academicians had broken the code was a secret of unparalleled importance, perhaps the most closely kept secret of the war, because the ability of the Bletchley machines to continue to successfully decode German messages depended on the Nazi high command's continuing ignorance of the fact that their unbreakable codes had been cracked.

Despite the importance of this work, early wartime bureaucracy and the thickets of secrecy surrounding the project threatened to cancel the incredible strategic advantage the 1943 Enigma breakthrough had handed the Allies. Turing appealed directly to Winston Churchill, who gave the project top priority. The codes continued to be cracked throughout the duration of the war, and in 1944 and 1945 the valuable information was disguised in the form of other kinds of intelligence, then relayed to British commanders in the Atlantic.

The tide of the critical U-boat conflict was turned, and the invasion of Europe became possible, largely because of Turing's successes with the naval version of the Enigma. The Germans never caught on, and Turing's esoteric work in metamathematics turned out to have dramatically practical applications after all. Because of the growing strategic significance of advanced cryptanalysis methods in the cold war era, the project continued to be held secret for decades after the war. After 1945, a very few people knew that Turing had done something important for the war effort but nobody knew exactly what it was, because he still wasn't allowed to allude to it.

His role at Bletchley wasn't Turing's only wartime contribution. He was sent over to America, at a time when it was indeed dangerous to take a North Atlantic cruise, to share crucial aspects of British cryptanalytic progress with American intelligence and to lend his intellectual assistance to several American war-related scientific projects.

It was during this American visit that Turing picked up practical knowledge of electronics. Turing had first become acquainted with what were then called "electronic valves" when he investigated the possibility of using the exotic vacuum-tube devices coming out of radar research to speed up the massive information-processing tasks needed by the Bletchley code-breakers. In America, Turing was involved in another hypersecret project, this time involving *voice* encryption—what the spy novels call "scramblers." Because of this work on the device that was code-named "Delilah," Turing learned his electronics from some of the best in the business—the engineers at Bell Laboratories in New York (including one named Claude Shannon, a prodigy of a different kind, who will enter this story again).

By the end of the war, the knowledge that electronic technology could be used to speed up logical switching circuits, and the possibility of building working models of Turing's universal machines, led His Majesty's government to once again support a scheme to construct an automatic calculating device. This time, it was not called the "Analytical Engine," but the "Automatic Computing Engine"—or *ACE*, as it became known. At the end of World War II, despite the work in America of Mauchly and Eckert (ENIAC's inventors), the British were in an excellent position to win the race to build the first true electronic digital computer. But unfortunately for Alan Turing, postwar computer research in Britain was not pursued as aggressively and on the same scale as the American effort.

Turing, of course, was in the thick of the postwar computer development effort, but not at the center, and certainly not in control. As it turned out, his heroic and secret war work helped to make him the victim of scientific politics, not their master. His reports on the hardware and software designs for ACE were ambitious, and if the machine he originally envisioned had been constructed as soon as it was designed, it would have put the ENIAC to shame.

While a succession of other men took over the direction of the computer projects at the National Physical Laboratory and at the University of Manchester, Turing hovered at the periphery of the political power while he put his mind to the actual construction of one of his long-imaginary universal machines. In this he was hampered by the attitude prevalent among his peers that upper-middle-class Cambridge theoreticians simply did not get their hands dirty with "engineering." But rigid conformity to social standards was not Alan's strong point. He forged ahead with what he knew was important—the development of a science of software.

Turing's ideas about the proper approach to computer design stressed the need to build computing capabilities into the programs, not the hardware. He was particularly interested in the programming operations—or "coding," as it was already coming to be called—by which truly interesting mathematical operations, and possibly "thinking" itself, eventually might be simulated by an electronic computer. And while Turing's first attempts at writing programming languages would be considered crude by today's standards, his ideas were far more advanced than the state of the hardware then available.

While his colleagues and the American teams scrambled to put together the most elementary models of electronic digital computers, Turing was already looking far beyond the clumsy contraptions constructed in the late forties and early fifties. His public talks and private conversations indicated a strong belief that the cost of electronic technology would drop while its power as a medium for computation would increase in the coming decades. He also believed that the capabilities of these devices would quickly extend beyond their original purposes.

Programs for doubling numbers or extracting square roots or breaking codes are handy tools, but Turing was aware that calculation was only one of the kinds of formal systems that could be imitated by a computational device. In particular, he saw how the simple "instruction tables" of his theoretical machines could become elements of a powerful grammar that the machines could use to modify their own operations.

One innovation of Turing's stemmed from the fact that computers based on Boolean logic operate only on input that is in the form of binary numbers (i.e., numbers expressed in powers of two, using only two symbols), while humans are used to writing numbers in the decimal system (in which numbers are expressed in powers of ten, using ten symbols). Turing was involved in the writing of instruction tables that *automatically* converted human-written decimals to machine-readable binary digits. If basic operations like addition, multiplication, and decimal-to-binary conversion could be fed to the machine in terms of instruction tables, Turing saw that it would be possible to build up *hierarchies* of such tables. The programmer would no longer have to worry about writing each and every operational instruction, step by repetitive step, and would thus be freed to write programs for more complex operations.

Turing wrote a proposal shortly after the end of the war, in which he discussed both the hardware and "coding" principles of a working model of his long-hypothetical machines. He foresaw that the creation of these instruction tables would become particularly critical parts of the entire process, for he recognized that the ultimate capabilities of computers would not always be strictly limited by engineering considerations, but by considerations of what was not yet known as "software."

Turing not only anticipated the fact that software engineering would end up being more difficult and time-consuming than hardware engineering, but anticipated the importance of what came to be called "debugging": [5]

> Instruction tables will have to be made up by mathematicians with computing experience and perhaps a certain puzzle-solving ability. There will probably be a good deal of work of this kind to be done, for every known process has got to be translated into instruction table form at some stage. This work will go on whilst the machine is being built, in order to avoid some of the delay between the delivery of the machine and the production of results. Delay there must be, due to the virtually invisible snags, for up to a point it is better to let the snags be there than to spend such time in design that there are none (how many decades would this course take?). This process of constructing instruction tables should be very fascinating. There need be no real danger of it ever becoming a drudge, for any processes that are quite mechanical may be turned over to the machine itself.

Except for the almost equally advanced ideas of a German inventor by the name of Konrad Zuse, which were long unknown to British and American scientists, Turing's postwar writings about the logical complexities and mathematical challenges inherent in the construction of instruction tables were the first significant steps in the art and science of computer programming. Turing was fascinated with the intricacies of creating coded instruction tables, but he was also interested in what might be done with a truly sophisticated programming language. His original work in metamathematical formalism had stemmed from his attempt to connect the processes of human thought to the structure of formal systems, and Turing was still intrigued by the possibility that automatic formal systems—computers—might one day emulate aspects of human reasoning.

The most profound questions Turing raised concerning the capabilities of universal machines were centered around this hypothesized future ability of computing engines to simulate human thought. If machinery might someday help in creating its own programming, would machinery ever be capable, even in principle, of performing activities that resembled human thought? His 1936 paper was published in a mathematical journal, but it eventually created the foundation of a whole new field of investigation beyond the horizons of mathematics—computer science. In 1950, Turing published another article that was to have profound impact; this piece, more simply titled "Computing Machinery and Intelligence," was published in the philosophical journal *Mind*.[6] In relatively few words, using no tools more esoteric than common sense, and absolutely no mathematical formulas, Turing provided the basis for the boldest subspecialty of computer science—the field of artificial intelligence.

Despite the simplicity of Turing's hypothetical machine, the formal description in the mathematics journal makes very heavy reading. The 1950 article, however, is worth reading by anyone interested in the issue of artificial intelligence. The very first sentence still sounds as direct and provocative as Turing undoubtedly intended it to be: "I propose to consider the question 'Can machines think?'"

In typical Turing style, he began his consideration of deep AI issues by describing—a game! He called this one "The Imitation Game," but history knows it as the "Turing test." Let us begin, he wrote, by putting aside the question of machine intelligence and considering a game played by three people—a man, a woman, and an interrogator of either gender, who is located in a room apart from the other two. The object of the game is for the interrogator to ask questions of the people in the other room, and to eventually identify which one is the man and which is the woman—on the basis of the answers alone. In order to disguise the appearance, voice, and other sensory clues from the players, the interrogation takes place over a teletype.

Turing then asks us to substitute a *machine* for one of the unknown players and make a new object for the game: This time, the interrogator is to guess, on the basis of the teletyped conversation, which inhabitant of the other room is a human being and which one is a machine. In describing how such a conversation might go, Turing quoted a brief "specimen" of such a dialogue:

Q: Please write me a sonnet on the subject of the Forth Bridge.
A: Count me out on this one. I could never write poetry.
Q: Add 34957 to 70764.
A: (pause about 30 seconds and then give as answer) 105621.
Q: Do you play chess?
A: Yes.

Q: I have K at my K1, and no other pieces. You have only K at
      K6 and R at R1. It is your move. What do you play?
A: (After a pause of 15 seconds) R-R8 mate.

Note that if this dialogue is with a machine, it is able to do faulty
arithmetic (34957 + 70764 does *not* equal 105621) and play decent
chess at the same time.

Having established his imitation game as the criterion for determining
whether or not a machine is intelligent, and before proceeding to consider
various objections to the idea of artificial intelligence, Turing explained
his own beliefs in the matter: [7]

> . . . I believe that in about fifty years' time it will be possible to
> program computers, with a storage capacity of about [ten billion bits]
> to make them play the imitation game so well that an average interroga-
> tor will not have more than 70 percent chance of making the right
> identification after five minutes of questioning. The original question,
> "Can machines think?" I believe to be too meaningless to deserve
> discussion. Nevertheless I believe that at the end of the century the
> use of words and general educated opinion will have altered so much
> that one will be able to speak of machines thinking without expecting
> it to be contradicted.

In the rest of the paper, Turing presented, then countered, a number
of principal objections to the possibility of artificial intelligence. The
titles Turing gave these objections reveal his whimsical streak: "The
Theological Objection," "The 'Heads in the Sand' Objection," "The
Mathematical Objection," "Lady Lovelace's Objection," "The Argument from Consciousness," "Arguments from Continuity in the Nervous
System," "The Argument from Informality of Behavior," and "The Argument from Extrasensory Perception."

In this paper, Turing made evident his knowledge of his intellectual
antecedents in this field by countering the objection raised by Ada in
her commentary, in which she stated the problem that is still cited by
most people in an argument about the possibility of machine intelligence:
"The Analytical Engine has no pretensions to *originate* anything. It
can do *whatever we know how to order it* to perform." Turing pointed
out that Ada might have spoken differently if she had seen, as he had,
evidence that electronic equipment could be made to exhibit a primitive
form of "learning," by which programs would be able to eventually

master tasks that had never been specifically programmed, but which emerged from trial-and-error techniques that had been programmed.

Turing's work in computing, mathematics, and other fields was cut short by his tragic death in June, 1954, at the age of forty-two. Besides being a genius, Turing was also a homosexual. During the early 1950s, following the defection of two homosexual spies to the Soviet Union, Great Britain was an especially harsh environment for anyone caught engaging in prohibited sexual acts—especially for someone who had something even more secret than radar or the atomic bomb in his head. Turing was arrested and convicted of "gross indecency," and sentenced to probation on the condition that he submit to humiliating and physically debilitating female hormone injections. Turing's war record was still too secret to even be mentioned in his defense.

Turing put up with the hormones and the public disgrace, and quietly began to break ground for another cycle of brilliant work in the mathematical foundations of biology—work that might have had even more momentous consequences, if it had been completed, than his work with computable numbers. For nearly two years after his arrest, during which time the homophobic and "national security" pressures grew even stronger, Turing worked with the ironic knowledge that he was being destroyed by the very government his wartime work had been instrumental in preserving. In June, 1954, Alan Turing lay down on his bed, took a bite from an apple, dipped it in cyanide, and bit again.[8]

Like Ada, Alan Turing's unconventionality was part of his undoing, and like her he saw the software possibilities that stretched far beyond the limits of the computing machinery available at the time. Like her, he died too young.

Other wartime research projects and other brilliant mathematicians were aware of Turing's work, particularly in the United States, where scientists were suddenly emerging into the nuclear age as figures of power. Military-sponsored research-and-development teams on both sides of the Atlantic continued to work on digital computers of their own. A few of these independent research efforts grew out of ballistics work. Others were connected with the effort to build the first nuclear fission and fusion bombs.

Over a hundred years had passed between Babbage and Turing. The computer age might have been delayed for decades longer if World War II had not provided top-notch engineering teams, virtually unlimited funds, and the will to apply scientific findings to real-world problems, at the exact point in the history of mathematics when the theory of computation made computers possible. While the idea undoubtedly would have resonated in later minds, whether or not the urgencies of war had forced the issue, the development of the computer was an inevitable engineering step once Turing explained computation.

When an equally, perhaps even more gifted thinker happened upon the same ideas Turing had been pursuing, it was no accident of history that Turing's theoretical insights were converted to workable machinery. A theory of computation is one very important step—but you simply cannot perform very sophisticated computations in a decently short interval if you are restricted to a box that chugs along a tape, erasing Os and writing Xs. The next step in both software and hardware history was precipitated by the thinking of another unique, probably indispensable figure in the history of programming—John von Neumann.

Turing had worked with von Neumann before the war, at Princeton's Institute for Advanced Study. Von Neumann wanted the young genius to stay on with him, as his protégé and assistant, but Turing returned to Cambridge. Von Neumann's profound understanding of the implications of Turing's work later became a significant factor in the convergence of different lines of research that led to the invention of the first digital computers.

It isn't often that the human race produces a polymath like von Neumann, then sets him to work in the middle of the biggest crisis in human history. Von Neumann was far more than an embellisher of Turing's ideas—he built the bridge between the abstractions of mathematicians and the practical concerns of people who were trying to create the first generation of electronic computers. He was a key member of the team who designed the software for the first electronic computer and who created the model for the physical architecture of computers. He also added elegance and power to Turing's first steps toward creating a true programming language.

CHAPTER FOUR

Johnny Builds Bombs and Johnny Builds Brains

If you asked ten thousand people to name the most influential thinker of the twentieth century, it is likely that not one of them would nominate John von Neumann. Few would even recognize his name. Despite his obscurity outside the communities of mathematicians and computer theorists, his thoughts had incalculable impact on human destiny. He died in 1957, but the fate of the human race still depends on how we and our descendants decide to use the technologies von Neumann's extraordinary mind made possible.

At the end of his life he was an American, and a power behind the scenes of American scientific policy and foreign policy. But that was only the last of several equally distinguished identities in different countries and fields of thought. Janos Neumann, known as "Jansci," was a prodigious young chemical engineer turned mathematician and logician in Hungary of the early 1920s. Johann von Neumann was one of the elite quantum physics revolutionaries in Göttingen, Germany, in the

67

late twenties. And from 1933 until his death, he was John von Neumann of Princeton, New Jersey; Los Alamos, New Mexico; and Washington, D.C., known to professors and Presidents as "Johnny."

Ada and Babbage could only dream of the day their device could be put to work. Turing was the tragic victim of political events before he could get his hands on a computer worth the name. Johnny, however, not only actually managed to get his machines built and use them to create the first working principles of software—but also ended up telling his government how to use the new technology. He was responsible for much more than the first boost in the accelerating American effort to develop computer technology.

A combination of many different scientific and political developments led to the invention of ENIAC. Electronic tube technology, Boolean logic, Turing-type computation, Babbage-Lovelace programming, and feedback-control theories were brought together because of the War Department's insatiable hunger for raw calculating power. John von Neumann was the only man who not only knew enough about the scientific issues but moved comfortably enough in the societies of Princeton and Los Alamos and Washington to grasp the threads and weave them together into an elegant and powerful design.

Von Neumann was a very important, probably indispensable, member of the Manhattan Project scientific team. Oppenheimer, Fermi, Teller, Bohr, Lawrence, and the other members of the most gifted scientific gathering of minds in history were as awed by Johnny's intellect as anyone else who ever met him. More impressively, they were as reliant on his mathematical judgment as anyone else. In that galactic cluster of world-class physicists, chemists, mathematicians, and engineers, it was a rare tribute that von Neumann was put in charge of the mathematical calculations upon which all their theories—and the functioning of their "gadget"—would depend.

As if his significant contributions to the development of the first nuclear weapons and the first computers were not enough for one man, he was also one of the original logicians who had posed the questions that Turing and Kurt Gödel answered in the 1930s. He was cofounder of the modern science of game theory (picking up where Babbage left off), one of the founders of operational research (also, curiously, advancing a field first explored by Babbage), an active participant in the creation of quantum physics, one of the first people to suggest analogies and differences between computer circuits and brain processes, and one of the first scientists since Turing to examine the relationship between the mathematics of code-making and the mystery of biological reproduction.

Von Neumann ended up as a key policy-maker in the fields of nuclear power, nuclear weapons, and intercontinental ballistic weaponry: he was the director of the Atomic Energy Commission and an influential member of the ICBM Committee. Generals and senators were lucky to get an appointment. Even when he was dying, the most powerful men in the world gathered around for a final consultation. According to Admiral Lewis Strauss, former chairman of the Atomic Energy Commission: "On one dramatic occasion near the end, there was a meeting at Walter Reed Hospital where, gathered around his bedside and attentive to his last words of advice and wisdom, were the Secretary of Defense and his Deputies, the Secretaries of the Army, Navy, and Air Force, and all the military Chiefs of Staff." [1]

John von Neumann's political views, undoubtedly rooted in his upper-class Hungarian past, were unequivocal and extreme, according to the public record as well as his biographers. He not only used his scientific expertise to hasten and accelerate the development of nuclear weapons and computer-guided missiles, but counseled military and political leaders to think about using these new American inventions against the USSR in a "preventive war." (In an article in *Life* magazine, published shortly after he died, von Neumann was quoted as saying: "If you say why not bomb them tomorrow, I say, why not today. If you say today at five o'clock, I say why not one o'clock.") [2]

In contrast to Turing, whom he knew from Turing's prewar stay at Princeton and from their wartime work, von Neumann was a sophisticated, worldly, and gregarious fellow, famous for the weekly cocktail parties he and his wife hosted during his tenure at Princeton's Institute for Advanced Study and up on the Mesa at Los Alamos. He had a substantial private income and an additional $10,000 a year from the Institute. He was widely known to have a huge repertoire of jokes in several languages, a vast knowledge of risqué limericks, and a casual manner of driving so recklessly that he demolished automobiles at regular intervals, always managing to emerge miraculously unscathed.

Despite his apparently charmed existence, von Neumann, like Ada Lovelace and Alan Turing, died relatively young. Lovelace died of cancer at thirty-six, Turing of cyanide at forty-two, and von Neumann of cancer at fifty-three. Like many other Los Alamos veterans, he may have been a victim of exposure to radiation during the early nuclear bomb tests. His death came as a shock to all who knew him as a vital, lively, peripatetic, seemingly invulnerable individual. Stanislaw Ulam, von Neumann's mathematical colleague and lifelong friend, in a memorial to Johnny published in a mathematics journal shortly after von Neumann's death, described his physical presence in loving detail: [3]

> Johnny's friends remember him in his characteristic poses: standing before a blackboard or discussing problems at home. Somehow, his gesture, smile, and the expression of the eyes always reflected the kind of thought or the nature of the problem under discussion. He was of middle size, quite slim as a very young man, then increasingly corpulent; moving about in small steps with considerable random acceleration, but never with great speed. A smile flashed on his face whenever a problem exhibited features of a logical or mathematical paradox. Quite independently of his liking for abstract wit, he had a strong appreciation (one might say almost a hunger) for the more earthy type of comedy and humor.

Everyone who knew him remembers to point out two things about von Neumann—how charming and personable he was, no matter what language he was speaking, and how much more intelligent than other human beings he always seemed to be, even in a crowd of near-geniuses. Among his friends, the standard joke about Johnny was that he wasn't actually human but was as skilled at imitating human beings as he was at everything else.

Born into an upper-class Hungarian Jewish family, Jansci was fluent in five or six languages before the age of ten, and he once told his collaborator Herman Goldstine that at the age of six he and his father often joked with each other in classical Greek. It was well known that he never forgot anything once he read it, and his ability to perform lightning-fast calculations was legendary.

One night in the middle of the summer of 1944, von Neumann encountered by happenstance a mathematician of past acquaintance in the Aberdeen, Maryland, train station. History might have been far different if one of their trains had been scheduled a few minutes earlier. That accidental meeting in Aberdeen presented von Neumann with a nearly completed approach to a problem the strategic significance of which he was uniquely equipped to understand, the details of which were complex and profound enough to attract his intellectual curiosity, the successful completion of which could be hastened by the use of his political clout.

Lieutenant Herman Goldstine, then associated with the U.S. Army Ordnance Ballistic Research Laboratory at Aberdeen, Maryland, didn't know anything about the other projects von Neumann was juggling at that time. But he knew that von Neumann's security clearance was miles above his and that he was a member of the Scientific Advisory Committee at the Ballistic Research Laboratory. So Goldstine happened to mention that an Army project at the Moore School of Engineering was soon going to produce a device capable of performing mathematical calculations at phenomenal speeds.

Years later, Goldstine remembered that he was understandably nervous upon meeting the world-famous mathematician on the railroad platform at the Aberdeen station. Goldstine recalled: [4]

> Fortunately for me, von Neumann was a warm friendly person who did his best to make people feel relaxed in his presence. The conversation soon turned to my work. When it became clear to von Neumann that I was concerned with the development of an electronic computer capable of 333 multiplications per second, the whole atmosphere of our conversation changed from one of relaxed good humor to one more like the oral examination for the doctor's degree in mathematics.

Because he had all-important reasons for wanting a fast automatic calculator, von Neumann asked for a demonstration. At the Moore School of Engineering, he met the gadget's inventors, Mauchly and Eckert, and the next years saw Johnny adding Aberdeen as a regular stop on his Princeton-D.C.-Los Alamos shuttle. Like everything else he turned his mind to, von Neumann immediately seemed to see more clearly than anyone else the future potential of what was then only a crude prototype. While the other principal creators of the first electronic computer were either mathematicians or electrical engineers, von Neumann was also a superb *logician*, which enabled him to understand what few others did—that these gadgets were in a class quite far beyond that of superfast calculating engines.

From those early meetings in 1944 to the eras of ENIAC, EDVAC, UNIVAC, MANIAC, and (yes) JOHNNIAC, the problem of assigning legal and historical credit to the inventors of the first electronic digital computers becomes a tangled affair in which easy explanations are impossible and many conflicts are still unresolved. Goldstine—the other man on the railroad platform with von Neumann—had his own version of the key events of early computer history. Mauchly and Eckert had a distinctly different point of view. There was the tale of Stibitz at Bell Labs. IBM's Thomas Watson, Senior, had yet another story. And a man in Iowa named Atanasoff eventually had the unexpected last laugh in a courtroom in 1973.

Monumental court cases have been fought over the issue of assigning credit for the invention of the modern computer, and even the legal decisions have been somewhat murky. Certainly it was a field in which a few people all over the world, working independently, reached similar conclusions. In the case of the ENIAC team, it was a case of several determined minds working together.

It isn't hard to envision von Neumann coming onto the scene after others had worked for years on the considerable engineering problems involved in building ENIAC (Electronic Numerical Integrator and Calculator), then dominating the voice of the group when they articulated their discoveries, not out of self-aggrandizement, but because he undoubtedly had the most elegant way of stating the conclusions that the group had arrived at, working in concert. Because of von Neumann's prominence in other fields, and the way his charm worked on journalists as well as generals, he was often described by the mass media as the *sole* inventor of key concepts like the all-important "stored program"— a credit he never claimed himself.

Although the matter of assigning credit for the earliest computer hardware is a tricky business, there is no denying von Neumann's central role in the history of software. His contributions to the science of computation in the late forties and early fifties were preceded by even earlier theoretical work that led to the notion of computation. He was one of the principal participants in both of the lines of thought that converged in the construction of ENIAC—mathematical logic and ballistics.

John von Neumann's role in the invention of computation began nearly twenty years before the ENIAC project. In the late 1920s, between his major contributions to quantum physics, logic, and game theory, young Johann von Neumann of Göttingen was one of the principal players in the international game of mathematical riddles that started with Boole seventy years prior and led to Turing's invention of the universal machine a decade later.

The impending collision of philosophy and mathematics that was becoming evident at the end of the nineteenth century made mathematicians extremely uncomfortable. Slippery metaphysical concepts like those associated with human thought might have appealed to minds like Boole's or Turing's. But to David Hilbert of Göttingen and others of the early 1900s, such vagueness was a grave danger to the future of an enterprise that intended to reduce all scientific laws to mathematical equations.

The logical and metamathematical foundations of more "pure" forms of mathematics, Hilbert insisted, could only be stated clearly in terms of numerical problems and precisely defined symbols and rules of operations. This was the doctrine of *formalism* that later spurred Turing to make his astonishing discovery about the capabilities of machines. Johann von Neumann, a student of Hilbert's, was one of the stars of the formalists. In itself, von Neumann's metamathematical achievement was remarkable. His work in formalism, however, was only part of what von Neumann achieved in several disparate fields, all in the same dazzling year.

In 1927, at the age of twenty-four, von Neumann published five papers that were instant hits in the academic world, and which still stand as monuments in three separate fields of thought. It was one of the most remarkable interdisciplinary triple plays in history. Three of his 1927 masterpieces were critical to the field of quantum physics. Another paper established the new field of game theory. The paper most directly related to the future of computation was about the relationship between formal logic systems and the limits of mathematics.

In this last 1927 paper, von Neumann demonstrated the necessity of proving that all mathematics was consistent, a critically important step toward establishing the theoretical bases for computation (although nobody yet knew that). This led, one year later, to a paper published by Hilbert that listed three unanswered questions about mathematics that he and von Neumann had determined to be the most important questions facing logicians and mathematics of the modern era.

The first of these questions asked whether or not mathematics was *complete*. Completeness, in the technical sense used by mathematicians, means that every true mathematical statement can be proven (i.e., is the last line of a valid proof).

The second question, the one that most concerned von Neumann, asked whether mathematics (or any other formal system) was *consistent*. Consistency in the technical sense means that there is no valid sequence of allowable steps (or "moves" or "states") that could prove an untrue statement to be true. If arithmetic was a consistent system, there would never be a way to prove that $1 + 1 = 3$.

The third question, the one that opened the side door to computation, asked whether or not mathematics was *decidable*. Decidability means that there is some definite method that is guaranteed to correctly determine whether any assertion is provable.

It didn't take long for a shocking answer to emerge in response to the first Hilbert–von Neumann question. In 1930, yet another young mathematician and logician, Kurt Gödel, showed that arithmetic cannot be complete, because there will always be at least one true assertion that cannot be proved. In the course of demonstrating this, Gödel crossed a crucial threshold between logic and mathematics when he showed that any formal system that is at least as rich as the number system (i.e., contains the mathematical operators + and =) can be expressed in terms of arithmetic. This means that no matter how complicated mathematics (or any other equally powerful formal system) becomes, it can always be expressed in terms of operations to be performed on numbers, and the parts of the system (whether or not they are inherently numerical) can be manipulated by rules of counting and comparing.

Von Neumann's and Hilbert's third question about the decidability of mathematics led Turing to his 1936 breakthrough. The "definite method" (of determining whether a mathematical assertion is provable) that was demanded by the decidability question was formulated by Alan Turing as a machine that could operate in definite steps on statements encoded as symbols on tape. Gödel had shown how numbers could represent the operations of a formal system, and Turing showed how the formal system could be described numerically to a machine equipped to decode such a description (e.g., translate the system's rules into the form "find a number n, such that . . . ", "n" being expressible as a string of ones and zeroes).

All of these theoretical questions were terribly important at the time they were formulated—to the few dozen people around the world who were equipped to understand their significance. But in 1930, the rest of the population had more important things to worry about than the hypothetical machines of the metamathematicians. Even those few who understood that universal machines could in fact be built were in no position to begin such a task. Making a digital computer was an engineering project that would require the kind of support that only a national government could afford.

John von Neumann was at the Institute for Advanced Study at Princeton by the time young Gödel and Turing came along. Although he was keenly aware of the latest developments in the "foundation crisis of mathematics" he had helped initiate in the late 1920s, von Neumann's restless intellect was attacking half a dozen new problems by the early 1930s. To Johnny, still in his twenties, the most important thing in life was to find "interesting problems."

In particular, he was interested in mathematical questions involving the phenomenon of turbulence, and the dynamics of explosions and implosions happened to be one area where such questions could be applied. He was also interested in new mathematical methods for modeling complex phenomena like global weather patterns or the passage of radiation through matter—methods that were powerful but required such enormous numbers of calculations that future progress in the field was severely limited by the human inability to calculate the results of the most interesting equations in a reasonable length of time.

Von Neumann seems to have had a kind of "Midas touch." The problems he tackled, no matter how abstruse and apparently obscure they might have seemed at the time, had a way of becoming very important a decade or two later. For example, he wrote a paper in the 1920s on the mathematics underlying economic strategies. A quarter of a century later it turned out to be a perfect solution to the problem of how airplanes should search for submarines (as well as one of the first triumphs of "operational research," one of the fields pioneered by Babbage).[5]

By the 1940s, von Neumann's expertise in the mathematics of hydrodynamic turbulence and the management of very large calculations took on unexpected importance because these two specialties were especially applicable to a new kind of explosion that was being cooked up by some of the old gang from Göttingen, now gathered in New Mexico. The designers of the first fission bomb knew that hellish mathematical problems in both areas had to be solved before any of the elegant equations of quantum physics could be transformed into the fireball of a nuclear detonation. As von Neumann already suspected, the mathematical work involved in designing nuclear and thermonuclear weapons created an avalanche of calculations.

The calculating power needed in the quest for thermonuclear weaponry ended up being one of the highest-priority uses for ENIAC—top-secret calculations for Los Alamos were the subject of the first official programs run on the device when it became operational—although the reason the electronic calculator had been commissioned in the first place was to generate the mathematical tables needed for properly aiming conventional artillery.

The ENIAC project was started under the auspices of the Army Ballistic Research Laboratory. Herman Goldstine, a historian of computation as well as one of the key participants, took the trouble to point out that the word *ballistics* is derived from the Latin *ballista*, the name of a large device for hurling missiles. Ballistics in the modern sense is the mathematical science of predicting the path of a projectile between the time it is launched and the moment it hits the target. Complex equations concerning moving bodies are complicated further by the adjustments necessary for winds of different velocities and for the variations in air resistance encountered by projectiles fired from very large guns as they travel through the atmosphere. The results of all possible distance, altitude, and weather calculations for guns of each specific size and muzzle velocity are given in "firing tables" which artillerymen consult as they set up a shot.

The application of mass-production techniques to weapons meant that new types of guns and shells were coming along at an unprecedented pace, making the ongoing production of firing tables no easy task. During World War I, such calculations were done by humans who were called "computers." But even then it was clear that new methods of organizing these large-scale calculations, and new kinds of mechanical calculators to help the work of the human computers, would be an increasingly important part of modern warfare.

In 1918 the Ballistics Branch of the Chief of Ordnance set up a special mathematical section at the Aberdeen Proving Ground in Maryland. One of the early recruits was the young Norbert Wiener, who was to feature prominently in another research tributary of the mainstream of ballistic technology—the automatic control of antiaircraft guns—and who was later to become one of the creators of the new computation-related discipline of *cybernetics.*

In the 1930s, both the Aberdeen laboratory and an associated group at the University of Pennsylvania's Moore School of Engineering obtained models of the automatic analog computer constructed by Vannevar Bush at MIT, a gigantic mechanical device known as the "differential analyzer." It was a marvelous aid to calculation, but it was far from being a digital computer, in either its design or its performance.

With the aid of these machines, the work of performing ballistic calculations was somewhat relieved. Before World War II, the machines were still second to the main resource—mathematics professors emeriti at the Moore School, who performed the calculations by hand, with the aid of hand-cranked mechanical calculators. Shades of Babbage's Cornish clergymen!

When war broke out, it was obvious that the institutions in charge of producing ballistic calculations for the several armed services needed expert help. It was for this reason that a mobilized mathematician, Lieutenant Herman Goldstine, reported for duty at Aberdeen in August, 1942, and was assigned to the task of streamlining ballistic computations. He soon found the Moore School facilities inadequate, and started to expand the staff of human "computers" by adding a large number of young women recruited from the Women's Army Corps to the small cadre of elderly ex-professors.

Goldstine's wife, Adele, herself a mathematician who was to play a prominent role in the programming of early computers (she and six other women were eventually assigned the task of programming the ENIAC), became involved with recruiting and teaching new staff members. Von Neumann's wife, Klara, performed a similar role at Los Alamos, both before and after electronic computing machines became available. The tradition of using women for such work was widespread—the equivalent roles in Britain's code-breaking efforts were played by hundreds of skilled calculators whom Turing and his colleagues called "girls" as well as "computers."

The expansion of the human computing staff at Aberdeen to nearly two hundred people, mostly WACs, was a stopgap measure. The calculation of firing tables was already out of hand. As soon as a new kind of gun, fuse, or shell became available for combat, a new table had to be calculated. The final product either was printed in a booklet that gunners kept in their pockets, or was mechanically encoded in special aiming apparatus called *automata*. (An entirely different mathematical research effort by Julian Bigelow, Warren Weaver, and Norbert Wiener was to concentrate on the characteristics of these automatic aiming machines.)

The answer to the firing table dilemma, as Goldstine was one of the first to recognize, was to commission the invention of an entirely new kind of mechanical calculating aid. The Vannevar Bush calculators were no longer the most efficient calculating devices. Faster machines, built on different principles, had been constructed by Dr. Howard Aiken and an IBM team at Harvard, and by a group led by a man named George Stibitz at Bell laboratories. But Goldstine knew that what they really needed at Aberdeen and the Moore School was an automatic calculator that was hundreds, even thousands of times faster than the fastest existing machines.

Such dreams of a supercalculator would have been akin to an Air Force officer wishing for a ten-thousand-mile-per-hour airplane, except for the fact that another new technology, one that only a few people even thought of applying to mathematical problems, looked as if it might make such a machine possible in theory, if only questionably probable in execution. Research in the young field of electronics had been uncovering all sorts of marvelous properties of the vacuum tube. Over in Great Britain, the whiz kids at Bletchley Park were using such devices in *Colossus*, their not-quite-computational code-breaking machine.

Until the war, electronic vacuum tubes had been used almost exclusively as amplifiers. But they could also be used as very fast switches. Since the rapid execution of a large number of on/off impulses is the hallmark of digital computation, and vacuum tubes could switch on and off as fast as a million times a second, electronic switching (as opposed to the mechanical switching of Vannevar Bush's machine) was an unbelievably good candidate for the key component of an ultrafast computing machine.

By 1943, unknown to Goldstine and almost all of his superiors, another, much higher-ranking scientist was also searching for an ultrafast computing machine. Goldstine beat the other fellow to it. Goldstine found Mauchly and Eckert in 1942. John von Neumann, and chance, found Goldstine in 1944.

John W. Mauchly and J. Presper Eckert have been properly credited with the invention of ENIAC, but before they implemented the key ideas of electronic digital computing machines, a man named Atanasoff in Iowa, in the 1930s, built small, crude, but functioning prototypes of electronic calculating machines. His name has not been as widely known, and his fortunes turned out differently from those of the other pioneers when computers grew from an exotic newborn technology to a powerful infant industry. But in 1973 a United States district court ruled that John Vincent Atanasoff invented the electronic digital computer.

It was a complicated decision, reached after years of litigation, and was not as clear-cut as it might have been if both sides did not have such strong cases. The core of the dispute centered around original work Atanasoff did in the 1930s, and the influence that his work later had on John Mauchly's design of ENIAC. Like the Hollerith-Billings story of the invention of punched-card data processing, simple explanations of where one man's ideas left off and another's began are difficult to reconstruct, at best.

Atanasoff was the last of the lone inventors in the field of computation; after him, such projects were too complicated for anything less than a team effort. Like Boole, Atanasoff was the recipient of one of those sudden inspirations that provided the solution to a problem he had been grappling with for years. A theoretical physicist teaching at Iowa State in the early 1930s, he came up against the same obstacle faced by other mathematicians and physicists of his era. The approaches to the most interesting ideas were blocked by the problems of performing large numbers of complex calculations.

By 1935, Atanasoff was in hot pursuit of a scheme to mechanize calculation. He was aware of Babbage's ideas, but he was an electronic hobbyist as well as a physicist, and entire technologies that didn't exist in Babbage's time were now showing great promise. Atanasoff was gradually convinced that an electronic computing machine was a good bet to pursue, but he had no idea how to go about designing one, and he wasn't sure how to design a machine without working out a method of programming it. In the late 1970s, Atanasoff told writer Katherine Fishman: [6]

> I commenced to go into torture. For the next two years my life was hard. I thought and thought about this. Every evening I would go off to my office in the physics building. One night in the winter of 1937 my whole body was in torment from trying to solve the problems of the machine. I got in my car and drove at high speeds for a long while so I could control my emotions. It was my habit to do this for a few miles: I could gain control of myself by concentrating on driving. But that night I was excessively tormented, and I kept on going until I had crossed the Mississippi River into Illinois and was 189 miles from where I started. I knew I had to quit; I saw a light, which turned out to be a roadhouse, and I went in. It was probably below zero outside, and I remember hanging up my heavy coat; I started to drink and commenced to warm up and realized I had control of myself.

Nearly forty years later, when he testified in the patent case concerning the invention of the electronic computer, Atanasoff recalled that he decided upon several design elements and principles that night in the roadhouse—including a binary system for encoding input and electronic tube technology for switching—that would transform his dream of an electronic calculator into a practical plan.

Atanasoff and a graduate assistant by the name of Clifford Berry obtained a research grant of $650 from the Iowa State University Research Council to build a prototype of their Atanasoff-Berry Calculator, known as the ABC. More grants followed when they completed the prototype in 1939. In 1940, at a meeting of the American Association for the Advancement of Science, Atanasoff met John Mauchly, who also had a kind of calculator to demonstrate, and who shared Atanasoff's conviction that powerful calculating machines would make possible significant scientific achievements that otherwise couldn't be approached.

The state of each inventor's mind at the time of their discussions in 1940 and 1941 was the crux of the legal and historical conflict. There is no dispute that John Mauchly had also devoted years of thought to the idea of automated calculation. Thirty-three years old when he met Atanasoff, Mauchly had worked his way through Johns Hopkins as a research assistant, which gave him extensive experience with procedures that involve detailed measurement and calculation. In 1933, as head of the physics department at Ursinus College near Philadelphia, he began to perform research in atmospheric electricity.

Mauchly was particularly interested in the long-disputed theory about the effects of sunspots on the earth's weather. There was no obvious connection between these huge storms on the sun and terrestrial weather conditions, but that did not prove that such a connection did not exist. In 1936, Mauchly arranged to have many parts of the government's voluminous meteorological records shipped back to his office at Ursinus. He intended to apply modern statistical analysis to the weather data in an attempt to correlate them with records of sunspot activity, hoping that this probe would reveal the previously undetected pattern.

As other mathematical meteorologists like von Neumann were also discovering, Mauchly found that any calculations involving data based on the weather quickly grew so complicated that it would take a lifetime to calculate all the equations generated from even the shortest periods of observation. So he found himself doing the same thing that ballistics experts did—hiring a lot of people with adding machines. A Depression-era agency, the National Youth Administration, helped Mauchly pay students fifty cents an hour to tabulate his weather data with hand calculators. Mauchly planned to obtain punched-card machines, once he got his crew to tackle the first part of the data. But when he watched a demonstration of the world's most advanced punched-card tabulator at the 1939 World's Fair, he realized that even scores of such machines in the hands of trained operators might take another decade to go through the weather data.

In 1939 and 1940, Mauchly read in scientific journals about a new measuring and counting system developed to assist cosmic-ray research. The part of the system that caught his eye was the fact that this new device, using electronic circuits, could count cosmic rays far faster than a dozen of the fastest punched-card tabulators. Cosmic rays can be detected at the rate of thousands per second, but all previous recorders failed to keep pace beyond 500 times a second. Mauchly tried making a few electronic circuits for himself, and began to see a way that they could be used for computation.

Mauchly took note of one circuit in particular that was developed by the cosmic-ray researchers—the *coincidence circuit*, in which a switch would be closed only when several signals arrived at exactly the same time, thus, in effect, rendering a decision. Would a machine capable of making electronic logical operations be possible via some variation of this circuit? Experimenting with his own vacuum-tube circuits, Mauchly speculated that there might also exist circuits used in other kinds of instruments that would enable him to build a machine to add, subtract, multiply, and divide. At this point, his speculations were more grandiose than his hand-wired prototypes, but the clues he had obtained from the cosmic ray researchers were enough to put Mauchly's weather-predicting machines on a collision course with a certain device the U.S. Army had in mind, one that had nothing to do with sunspots or the weather.

Mauchly brought a small analog device to the AAAS meeting where he met Atanasoff, and in June, 1941, he hitched a ride to visit Atanasoff in Ames, Iowa. Atanasoff demonstrated the ABC, Mauchly stayed for five days, and thirty-two years later a court decided that Mauchly's later invention of the ENIAC relied upon key ideas of Atanasoff's that were transferred from mind to mind those five days in June.

The 1973 legal decision (*Honeywell versus Sperry Rand*, U.S. District Court, District of Minnesota, Fourth Division) did not state that Mauchly stole anything, but did restore partial credit for the invention of the electronic computer to a man whose name had been nearly forgotten in all the publicity and honors heaped upon Mauchly and Eckert. After the ruling, Mauchly said: "I feel I got nothing out of the visit to Atanasoff except the royal shaft later." [7] On Mauchly's behalf, it must be noted that nobody has disputed the fact that the sheer scale and engineering audacity of ENIAC was far beyond the ABC, and that Mauchly was indeed on the right track at least as early as Atanasoff.

Part of the reason for ENIAC's success and ABC's obscurity must be attributed to the accidents of history. Legal issues aside, the historical momentum shifted to Mauchly later in the summer of 1941, when he signed up for an Army-sponsored electronics course at the Moore School of Engineering. His instructor, J. Presper Eckert, was an exceptionally bright Philadelphia blueblood twelve years younger than Mauchly. When Eckert, the electronics wizard, learned of Mauchly's plan to automate large-scale numerical calculations, a critical mass of idea-power was reached. They were in exactly the right place at the right time to cook up such an ambitious project.

Not long after thirty-four-year-old John Mauchly and twenty-two-year-old Pres Eckert started to sketch out a plan for an electronic computer, they became acquainted with Lieutenant Herman Goldstine, who enthusiastically joined their discussions, both as a mathematician and as liaison

The U.S. Army/Moore School ENIAC team. From left to right: J. Presper Eckert, Jr., Chief Engineer; Professor J. G. Brainerd, Supervisor; Sam Feltman, Chief Engineer for Ballistics, Ordnance Department; Captain H. H. Goldstine, Liaison Officer; Dr. J. W. Mauchly, Consulting Engineer; Dean Harold Pender, Moore School of Electrical Engineering, University of Pennsylvania; General G. M. Barnes, Chief of the Ordnance Research and Development Service; Colonel Paul N. Gillon, Chief, Research Branch of the Army Ordnance Research and Development Service. (Sperry Corporation. Courtesy of Eleutherian Mills-Hagley Foundation, Inc.)

officer between the Moore School and the Ballistic Research Laboratory. By the time he met them, Goldstine was sufficiently frustrated by the lack of ballistic calculating power that he was receptive to even a science-fiction story like the one presented to him by these two whiz kids.

As wild as it sounded as an engineering feat, Goldstine knew that an electronic device such as the one Mauchly and Eckert described to him had the potential to perform ballistic calculations over 1000 times faster than the best existing machine, the Aiken-IBM-Harvard-Navy device called the Mark I. But it would cost a lot of money to find out if they were right. Atanasoff and Berry built their prototypes for a total of $6500. These boys would need hundreds of thousands of dollars to lash together something so complicated and delicate that most electrical engineers of the time would swear it could never work.

Goldstine later explained the risks associated with attempting the proposed electronic calculator project: [8]

> . . . we should realize that the proposed machine turned out to contain over 17,000 tubes of 16 different types operating at a fundamental clock rate of 100,000 pulses per second. . . . once every 10 microseconds an error would occur if a single one of the 17,000 tubes operated incorrectly; this means that in a single second there were 1.7 billion . . . chances of a failure occurring . . . Man had never made an instrument capable of operating with this degree of fidelity or reliability, and this is why the undertaking was so risky a one and the accomplishment so great.

The two young would-be computer inventors at the Moore School, the mathematician-turned-lieutenant who found them, and their audacious plan for cutting through the calculation problem by creating the world's most complicated machine were the subject of a high-level meeting on April 9, 1943. Attending was one of the original founders of the military's mathematical research effort and President of the Institute for Advanced Study at Princeton, Oswald Veblen, as well as Colonel Leslie Simon, director of the Ballistic Research Laboratory, and Goldstine.

The moment when the United States War Department entered the age-old quest for a computing machine, and thus made the outcome inevitable, was recalled by Goldstine when he wrote, nearly thirty years later, that Veblen, "after listening for a short while to my presentation and teetering on the back legs of his chair brought the chair down with a crash, arose, and said, 'Simon, give Goldstine the money.' " [9] They got their money—eventually, as much as $400,000—and started building their machine.

ENIAC was monstrous—100 feet long, 10 feet high, 3 feet deep, weighing 30 tons—and hot enough to keep the room temperature up toward 120 degrees F while it shunted multivariable differential equations through its more than 17,000 tubes, 70,000 resistors, 10,000 capacitors, and 6000 hand-set switches. It used an enormous amount of power—the apocryphal story is that the lights of Philadelphia dimmed when it was plugged in.

When it was finally completed, ENIAC was too late to use in the war, but it certainly delivered what its inventors had promised: a ballistic calculation that would have taken twenty hours for a skilled human calculator could be accomplished by the machine in less than thirty seconds. For the first time, the trajectory of a shell could be calculated in less time than it took an actual shell to travel to its target. But firing tables were no longer the biggest boom on the block by the time ENIAC was completed. The first problem run on the machine, late in the winter of 1945, was a trial calculation for the hydrogen bomb then being designed.

After his first accidental meeting with Goldstine at Aberdeen, and the demonstration of a prototype of the ENIAC soon afterward, von Neumann joined the Moore School project as a special consultant. Johnny's genius for formal, systematic, logical thinking was applied to the logical properties of this huge maze of electronic circuits. The engineering problems were still formidable, but it was also becoming clear that the nonphysical component, the subtleties of setting up the machine's operations—the *coding*, as they began to call it—was equally difficult and important.

Until the transistor came along a few years later, ENIAC would represent the physical upper limit of what could be done with a large number of high-speed switches. In 1945, the most promising approach to greater computing power was in improving the logical structure of the machine. And von Neumann was perhaps the one man west of Bletchley Park equipped to understand the logical attributes of the first digital computer.

Part of the reason ENIAC was able to operate so fast was that the routes followed by the electronic impulses were wired into the machine. This electronic routing was the materialization of the machine's instructions for transforming the input data into the solution. Many different kinds of equations could be solved, and the performance of a calculation

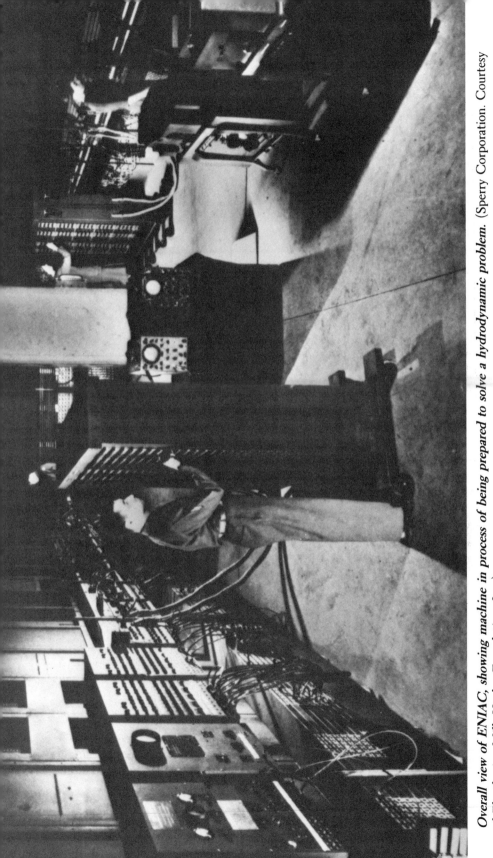

Overall view of ENIAC, showing machine in process of being prepared to solve a hydrodynamic problem. (Sperry Corporation. Courtesy of Eleutherian Mills-Hagley Foundation, Inc.)

could be altered by the outcome of subproblems, but ENIAC was no-where near as flexible as Babbage's Analytical Engine, which could be reprogrammed to solve a different set of equations, not by altering the machine itself, but by altering the sequence of input cards.

What Mauchly and Eckert gained in calculating power and speed, they paid for in overall flexibility. The gargantuan electronic machine had to be set up for solving each separate problem by changing the configuration of a huge telephone-like switchboard, a procedure that could take days. The origins of the device as a ballistics project were partially responsible for this inflexibility. It was not the intention of the Moore School engineers to build a universal machine. Their contract quite clearly specified that they create an altogether new kind of trajectory calculator.

Especially after von Neumann joined the team, they realized that what they were constructing would not only become the ultimate mathematical calculator, but the first, necessarily imperfect prototype of a whole new category of machine. Before ENIAC was completed, its designers were already planning a successor. Von Neumann, especially, began to realize that what they were talking about was a *general-purpose machine*, one that was by its nature particularly well suited to function as an extension of the human mind.

If one thing was sacred to Johnny, it was the power of human thought to penetrate the mysteries of the universe, and the will of human beings to apply that knowledge to practical ends. He had other things on his own mind at the time—from the secrets of H-bomb design to the structure of logic machines—but he appeared to be most keen on the idea that these devices might evolve into some kind of intellectual extension. How much more might a thinker like himself accomplish with the aid of such a machine? One biographer put it this way: [10]

Von Neumann's enthusiasm in 1944 and 1945 had first been generated by the challenge of improving the general-purpose computer. He had been the proponent of using the latest in computing machines in the atomic bomb project, but he realized that for the impending hydrogen bomb project still better and faster machines were needed. On the theoretical level he was intrigued by the fact that there appeared to be organizational parallels between the brain and computers and that these parallels might lead to formal-logical theories encompassing both computers and brains; moreover, the logical theories would constitute interesting abstract logics in their own right. He

FUNCTION TABLE A

Army ordnance personnel preparing ENIAC for a ballistics calculation by setting the switches. This tedious process was replaced by EDVAC's stored-program scheme. (Sperry Corporation. Courtesy of Eleutherian Mills-Hagley Foundation, Inc.)

was cautious in assuming similarity between a computer and the awe-some functioning of the human brain, especially as in 1944 he had little preparation in physiology. Rather he regarded the computer as a technical device functioning as an extension of its user; it would lead to an aggrandizement of the human brain, and von Neumann wanted to push this aggrandizement as far and as fast as possible.

There is no dispute that Mauchly, Eckert, Goldstine, and von Neumann worked together as a team during this crucial gestation period of computer technology. The team split up in 1946, however, so the matter of accrediting specific ideas has become a sticky one. Memoranda were written, as they are on any project, without the least expectation that years later they would be regarded as historical or legal documents. Technology was moving too fast for the traditional process of peer review and publication: the two most important documents from these early days were titled "First Draft . . ." and "Preliminary Report . . ."

By the time they got around to sketching the design for the next electronic computer, the four main ENIAC designers agreed that the goal was to design a machine that would use the same hardware technology in a much more efficient way. The next step, the invention of *stored programming*, is where the accreditation controversy comes in. At the end of June, 1945, the ENIAC team prepared a proposal in the form of a "First Draft of a Report on the Electronic Discrete Variable Calculator" (EDVAC). It was signed by von Neumann, but reflected the conclusions of the group. Goldstine later said of this: "It has been said by some that von Neumann did not give credits in his *First Draft* to others. The reason for this was that the document was intended by von Neumann as a working paper for use in clarifying and coordinating the thinking of the group and was not intended for publication." [11] (Mauchly and Eckert, however, took a less benign view of von Neumann's intentions.) The most significant innovations articulated in this paper involved the logical aspects of coding, as well as dealing with the engineering of the physical device that was to follow the coded instructions.

Creating the coded instructions for a new computation on ENIAC was nowhere near as time-consuming as carrying out the calculation by hand. Once the code for the instructions needed to carry out the calculation had been drawn up, all that had to be done to perform that computation on any set of input data was to properly configure the machine to conform with those instructions. The calculation, which formerly took up the most time, had become trivial, but a new bottleneck was created with the resetting of switches, a process that took an unreasonable amount of time compared with the length of time it would take to run the calculation.

Resetting the switches was the most worrisome bottleneck, but not the only one. The amount of time it took for the instructions to make use of the data, although greatly reduced from the era of manual calculation, was also significant—in ballistics, the ultimate goal of automating calculation was to be able to predict the path of a missile *before* it landed, not days or hours or even just minutes later. If only there was a more direct way for the different sets of instructions—the inflexible, slow-to-change component of the computing system—to interact with the data stored in the electronic memory, the more quickly accessible component of computation. The solution, as von Neumann and colleagues formulated it, was an engineering innovation based on a *logical* breakthrough.

The now-famous "First Draft" described the logical properties of a true general-purpose electronic digital computer. In one key passage, the EDVAC draft pointed out something that Babbage, if not Turing, had overlooked: "The device requires a considerable memory. While it appeared that various parts of this memory have to perform functions which differ somewhat in their nature and considerably in their purpose, it is nevertheless tempting to treat the entire memory as one organ." [12] In other words, a general-purpose computer should be able to store instructions in its internal memory, along with data.

What used to be a complex configuration of switchboard settings could be symbolized by the programmer in the form of a number and read by the computer as the location of an instruction stored in memory, an instruction that would automatically be applied to specified data that was also stored in the memory. This meant that programs could call up other programs, and even modify other programs, without intervention by the human operator. Suddenly, with this simple change, true information processing became possible.

This is the kernel of the concept of stored programming, and although the ENIAC veterans were officially the first to describe an electronic computing device in such terms, it should be noted that the abstract version of exactly the same idea was proposed in Alan Turing's 1936 paper in the form of the single tape of the universal Turing machine. And at the same time the Pennsylvania group was putting together the EDVAC report, Turing was thinking again about the concept of stored programs: [13]

So the spring of 1945 saw the ENIAC team on the one hand, and Alan Turing on the other, arrive naturally at the idea of constructing a universal machine with a single "tape.". . .

But when Alan Turing spoke of "building a brain," he was working and thinking alone in his spare time, pottering around in a British back garden shed with a few pieces of equipment grudgingly conceded by the secret service. He was not being asked to provide the solution to numerical problems such as those von Neumann was engaged upon; he had been thinking for himself. He had simply put together things that no one had put together before: his one tape universal machine, the knowledge that large-scale electronic pulse technology could work, and the experience of turning cryptanalytic thought into "definite methods" and "mechanical processes." Since 1939 he had been concerned with little but symbols, states, and instruction tables—and with the problem of embodying these as effectively as possible in concrete forms.

With the EDVAC design, ballistics calculators took the first step toward general-purpose computers, and it became clear to a few people that such devices would surely evolve into something far more powerful. The kind of uses the inventors envisioned for the future of their technology was cause for one of several major theoretical disagreements that were to surface soon thereafter among the four ENIAC principals. Von Neumann and Goldstine saw the opportunity to build an incredibly powerful research tool for scientists and mathematicians. Mauchly and Eckert were already thinking of business and government applications outside military or research institutions.

The first calculation run on ENIAC in December, 1945, six months after the "First Draft," was a problem posed by scientists from Los Alamos Laboratories. ENIAC was formally dedicated in February, 1946. By then, the patriotic solidarity enforced on the research team by wartime conditions had faded away. Von Neumann was enthusiastic about the military and scientific future of the computer-building enterprise, but the two young men who had dreamed up the computer project before the big brass stepped in were getting other ideas about how their brainchild ought to mature. The tensions between institutions, people, and ideas mounted until Mauchly and Eckert left the Moore School on March 31, 1946, over a dispute with the university concerning patent rights to ENIAC. They founded their own group shortly thereafter, eventually naming it *The Eckert-Mauchly Computer Corporation*.

When Mauchly and Eckert later suggested that they were, in fact, the sole originators of the main ideas of the EDVAC report, they were, in Goldstine's phrase, "strenuously opposed" by Goldstine and von Neumann. The split turned into a lifelong feud. Goldstine, writing in 1972 from his admittedly partial perspective, was unequivocal in pointing out von Neumann's contributions: [14]

> First, his entire summary as a unit constitutes a major contribution and had a profound impact not only on the EDVAC but also served as a model for virtually all future studies of logical design. Second, in that report he introduced a logical notation adapted from one of McCulloch and Pitts, who used it in a study of the nervous sytem. This notation became widely used, and is still, in modified form, an important and indeed essential way for describing pictorially how computer circuits behave from a logical point of view.
>
> Third, in the famous report he proposed a repertoire of instructions for the EDVAC, and in a subsequent letter he worked out a detailed programming for a *sort and merge* routine. This represents a milestone, since it is the first elucidation of the now famous stored program concept together with a completely worked-out illustration.
>
> Fourth, he set forth clearly the serial mode of operation of the modern computer, i.e., one instruction at a time is inspected and then executed. This is in sharp distinction to the parallel operation of the ENIAC in which many things are simultaneously performed.

While Mauchly and Eckert set forth to establish the commercial applications of computer technology, Goldstine, von Neumann, and another mathematician by the name of Arthur Burks put together a proposal and presented it to the Institute for Advanced Study at Princeton, the Radio Corporation of America, and the Army Ordnance Department, requesting one million dollars to build an advanced electronic digital computer. Once again, some of the thinking in this report was an extension of the group creations of the ENIAC project. But this "Preliminary Discussion," unquestionably dominated by von Neumann, also went boldly beyond the EDVAC conception as it was stated in the "First Draft."

Although the latest proposal was aimed at the construction of a machine that would be more sophisticated than EDVAC, the authors went much farther than describing a particular machine. They very strongly suggested that their specification should be the general plan for the logical structure and fundamental method of operation of all future computers. They were right: It took almost forty years, until the 1980s, until anyone made a serious attempt to build "non–von Neumann machines."

"Preliminary Discussion of the Logical Design of an Electronic Computing Instrument," which has since been recognized as the founding document of the modern science of electronic computer design, was submitted on June 28, 1946, but was available only in the form of mimeographed copies of the original report to the Ordnance Department until 1962, when a condensed version was published in *Datamation* magazine.[15] The primary contributions of this document were related to the logical use of the memory mechanism and the overall plan of what has come to be known as the "logical architecture." One aspect of this architecture was the ingenious way data and instructions were made to be changeable during the course of a computation without requiring direct intervention by the human operator.

This changeability was accomplished by treating numerical data as "values" that could be assigned to specific locations in memory. The basic memory component of an EDVAC-type computer used collections of memory elements known as "registers" to store numerical values in the form of series of on/off impulses. Each of these numbers was assigned an "address" in the memory, and any address could contain either data or an instruction. In this way, specific data and instructions could be located when needed by the control unit. One result of this was that a particular piece of data could actually be a variable—like the x in algebra—that could be changed independently by having the results of an operation stored at the appropriate address, or by telling the computer to perform an operation on whatever was found at that location.

One of the characteristics of any series of computation instructions is a reference to data: when the instructions tell the machine how to perform a calculation, they have to specify what data to plug into the calculation. By making the reference to data a reference to the contents of a specific memory location, instead of a reference to a specific number, it became possible for the data to change during the course of a computation, according to the results of earlier steps. It is in this way that the numbers stored in the memory can become symbolic of quantities other than just the numerical value, in the same way that algebra enables one to manipulate symbols like x and y without specifying the values.

It is easier to visualize the logic of this schema if you think of the memory addresses as something akin to numbered cubbyholes or post-office boxes—each address is nothing but a place to find a message. The addresses serve as easily located containers for the (changeable) values (the "messages") to be found inside them. Box #1, for example, might contain a number; box #2 might contain another number; box

#3 might contain instructions for an arithmetic operation to be per-
formed on the numbers found in boxes #1 and #2; box #4 might
contain the result of the operation specified in box #3. The numbers
in the first two boxes might be fixed numbers, or they might be variables,
the values of which might depend on the result of other operations.

By putting both the instructions and the raw data inside the same
memory, it became possible to perform computations much faster than
with ENIAC, but it also made it necessary to devise a way to clearly
indicate to the machine that some specific addresses contain instructions
and other addresses contain numbers for those instructions to operate
on.

In the "First Draft," von Neumann specified that each instruction
should be designated in the coding of a program by a number that
begins with the digit 1, and each of the numbers (data) should begin
with the digit 0. The "Preliminary Report" expanded the means of
distinguishing instructions from data by stating that computers would
keep these two categories of information separate by operating during
two different *time cycles*, as well.

All the instructions are executed according to a timing scheme based
on the ticking of a built-in clock. The "instruction" cycles and "execution" cycles alternate: On "tick," the machine's control unit interprets
numbers brought to it as instructions, and prepares to execute the operations specified by the instructions on "tock," when the "execution"
cycle begins and the control unit interprets input as data to operate
upon.

The plan for this new category of general-purpose computer not only
specified a timing scheme but set down what has become known as
the "architecture" of the computer—the division of logical functions
among physical components. The scheme had similarities to both Babbage's and Turing's models. All such machines, the authors of the "Preliminary Report" declared, must have a unit where arithmetic and logical
operations can be performed (the processing unit where actual calculation
takes place, equivalent to Babbage's "mill"), a unit where instructions
and data for the current problem can be stored (like Babbage's "store,"
a kind of temporary memory device), a unit that executes the instructions
according to the specified sequential order (like the "read/write head"
of Turing's theoretical machine), and a unit where the human operator
can enter raw information or see the computed output (what we now
call "input-output devices").

Any machine that adheres to these principles—no matter what physical

From left to right: Julian Bigelow, Herman H. Goldstine, J. Robert Oppenheimer, and John von Neumann, standing in front of the Institute for Advanced Study computer. Oppenheimer had been von Neumann's superior as director of the Manhattan Project. Bigelow had been a protégé of Norbert Wiener's. (Courtesy of the Institute for Advanced Study.)

technology is used to implement these logical functions—is an example of what has become known as "the von Neumann architecture." It doesn't matter whether you build such a machine out of gears and springs, vacuum tubes, or transistors, as long as its operations follow this logical sequence. This theoretical template was first implemented in the United States at the Institute for Advanced Study. Modified copies of the IAS machine were made for the Rand Corporation, an Air Force spinoff "think tank" that was responsible for keeping track of targets for the nation's new but fast-growing nuclear armory, and for the Los Alamos Laboratory. Against von Neumann's mild objections, the Rand machine was dubbed JOHNNIAC. The Los Alamos machine assigned to nuclear weapons-related calculations was given the strangely uneuphemistic name of MANIAC.

(Neither EDVAC, the IAS machine, the Los Alamos, nor the Rand machine was the first operational example of a fully functioning stored-program computer. British computer builders, who had been pursuing parallel research and who were aware of von Neumann's ideas, beat the Americans when it came to constructing a machine based on the logical principles enunciated by von Neumann. The first machine that was binary, serial, and used stored-program memory was the EDSAC— the Electronic Delay Storage Automatic Calculator, built at the University Mathematical Laboratory, University of Cambridge, England.)

In a von Neumann machine, the arithmetic and logic unit is where the basic operations of the system are wired in. All other instructions are constructed out of various combinations of these fundamentals. It is possible, in principle, to build a device of this type with very few, extremely simple, built-in operations. Addition, for example, could be performed over and over again whenever a multiplication operation is requested by a program. In fact, the only two operations that are absolutely necessary are "not" and "and." The problem with using a few very simple hardwired operations and proportionally complex software structures built from them is that it slows down the operations of the computer: Because instructions are executed one at a time ("serially") as the internal clock ticks, the number of basic instructions in a program dictates how long it takes a computer to run that program.

The control unit specified by the "Preliminary Report"—the component that supervises the execution of instructions—was the materialization of the formal logical device created by Emil L. Post and Turing, who had proved that it was possible to devise codes in terms of numbers that could cause a machine to solve any problem that was clearly statable. This is where the symbol meets the signal, where the sequences of on and off impulses in the circuits, the Xs and Os on the cells of the endless tape, the strings of numbers in the programmer's code, marry the human-created computation to the machine that computes.

The input-output devices were the parts of the system that were to advance the most slowly while the electronic switch-based memory, arithmetic, and control components ascended through orders of magnitude. For over a decade after ENIAC, punched cards were the main input devices, and for over two decades, teletype machines were the most common output devices.

The possibility of future breakthroughs in this area and their implications were not overlooked. In a memorandum written in November, 1945, concerning one of the early proposals for the IAS machine, von

Neumann anticipated the possibility of creating a more visually oriented output device: [16]

> In many cases the output really desired is not digital (presumably printed) but pictorial (graphed). In such situations the machine should graph it directly, especially because graphing can be done electronically and hence more quickly than printing. The natural output in such a case is an oscilloscope, i.e., a picture on its fluorescent screen. In some cases these pictures are wanted for permanent storage . . . in others only visual inspection is desired. Both alternatives should be provided for.

But a personal interactive computer, helpful as such a device might be to a mind such as von Neumann's, was not an interesting enough problem. After solving interesting problems about the processes that take place in the heart of stars, a scientific-technological tour de force that also became a historical point of no return when the scientists' employers demonstrated their creation at Hiroshima, and then solving another set of problems concerned with the creation of computing machinery, all the while pontificating about the most potent aspects of foreign policy to the leaders of the most powerful nation in history, John von Neumann was aiming for nothing less than the biggest secret of all. In the late 1940s and early 1950s, the most interesting scientific question of the day was "what is life?"

To someone who had been at Alamogordo and the Moore School, it would not have been too farfetched to believe that the next intellectual conquest might bring the secret of physical immortality within reach. Certainly he would never know whether he could truly resolve the most awesome of nature's mysteries unless he set his mind to decoding the secret of life. And that he did. Characteristically, von Neumann focused on the aspect of the mystery of life that appealed to his dearest instincts and most powerful capacities—the pure, logical, mathematical underpinnings of nature's code. He was particularly interested in the logical properties of the theoretical devices known as *automata*, of which Turing's machine was an example.

Von Neumann was especially drawn to the idea of *self-reproducing* automata—mathematical patterns in space and time that had the property of being able to reproduce themselves. He was able to draw on his knowledge of computers, his growing understanding of neurophysiology and biology, and make particularly good use of his deep understanding of logic, because he saw self-reproducing automata as essentially logical beasts. The way the task was accomplished by living organisms of the type found on earth was only one way it could be done. In principle, the task could be done by a machine that could follow a plan, because the plan, and not the mechanism that carried it out, was the part of the system with the special, heretofore mysterious property that distinguished life from nonliving matter.

Von Neumann approached "cellular automata" on an abstract level, just as Turing did with his first machines. As early as 1948, he showed that any self-replicating system must have raw materials, a program that provides instructions, an automaton that follows the instructions and arranges the symbols in the cells of a Turing-type machine, a system for duplicating instructions, and a supervisory unit—which turned out to be an excellent description of the DNA direction of protein synthesis in living cells.

Another thing that interested Johnny was the gamelike aspect of the world. Accordingly, he thought about the way his self-reproducing automaton was like a game: [17]

> Making use of work done by his colleague Stanislav Ulam, von Neumann was able to refine his calculations and make them more generally applicable. Von Neumann's mental experiment, which we can easily present in the form of a game, makes use of a homogeneous space subdivided into cells. We can think of these cells as squares on a playing board. A finite number of states—e.g., empty, occupied, or occupied by a specific color—is assigned to each square. At the same time, a neighborhood is defined for each cell. This neighborhood can consist of either the four orthogonally bordering cells or the eight orthogonally and diagonally bordering cells. In the space divided up this way, transition rules are then applied simultaneously to each cell. The transition any particular cell undergoes will depend on its state and on the states of its neighbors. Von Neumann was able to prove that a configuration of about 200,000 cells, each with 29 different possible states and each placed in a neighborhood of 4 orthogonally adjacent squares, could meet all the requirements of a self-reproducing automaton. The large number of elements was necessary because von Neumann's model was also designed to simulate a Turing machine. Von Neumann's machine can, theoretically, perform any mathematical operation.

In 1950, when it was evident to all that the engineering phase of computer technology was accomplishing impressive tasks, von Neumann postulated one such system in terms of a factory that contains within it the machinery and the detailed blueprints for making identical factories (and identical blueprints) from raw materials to be provided to it. Take a step up from that in complexity, and the details can include a specification for subsystems that find raw materials for the factory from the environment, with no further human intervention.

If one fantasizes one step farther on the complexity spectrum, the instructions and capabilities could specify factories capable of building spaceships to send more factories to other planets, where the raw materials found would be shaped into more factory-spaceship-launchpad systems, and if you could build factories that would each build *two* or more such complexes, you could have a counterforce to the general disorderly trend of the cosmos, in the form of a (mindless?) horde of factory-building factories, munching outward through the galaxies like an anti-entropic swarm of logical locusts.

While it definitely sounds like a science-fiction story, and many would add that it could be interpreted to be an idea of such inhuman coldness as to be termed "fiendish," such scenarios are legitimate topics in the field of automata, and are still known as "von Neumann machines" (as distinguished from "the von Neumann machine," the logical architecture he created for digital computers).

Von Neumann died in 1957, before he could achieve a breakthrough in the field of automata. Like Ada, he died of cancer, and like Ada, he was said to have suffered terribly, as much from the loss of his intellectual faculties as from the pain. But the world he left behind him was powerfully rearranged by what he had accomplished before he failed to solve his last, perhaps most interesting problem.

CHAPTER FIVE

Ex-Prodigies and Antiaircraft Guns

Today, when molecular biologists talk about the "coding" of the DNA molecule, cognitive scientists discuss the "software of the brain," and behavioral psychologists write about "reprogramming old habits," they are all making use of a scientific metaphor that emerged from the technology of computation, but which has come to encompass much more than the mechanics of calculating devices. *Cybernetics*, the study of communication and control in physical and biological systems, was born when yet another unusual mind was drawn into the software quest through the circumstances of war.

Because of the discoveries of Norbert Wiener and his colleagues, discoveries that were precipitated by the wartime need for a specific kind of calculating engine, software has come to mean much more than the instructions that enable a digital computer to accomplish different tasks. From the secrets of life to the ultimate fate of the universe, the principles of communication and control have been successfully applied to the most important scientific puzzles of our age. These principles were discovered through a strange concatenation of events, and the people who were involved in those events were no less unusual than the software patriarchs who preceded them.

Eccentrics and prodigies of both the blissful and agonized varieties dominated the early history of computation. Ada Lovelace, George Boole, John von Neumann, Alan Turing, and Presper Eckert were all in their early twenties or younger when they did their most important work. All except Eckert were also more than a little bizarre. But for raw prodigy combined with sheer imaginative eccentricity, Norbert Wiener, helmsman of the cybernetic movement, stands out even in this not-so-ordinary crowd.

Norbert's father, a Harvard professor who was a colorful character in his own right, had definite opinions about education, and publicly declared his intention to mold his young son's mind. Norbert was to become a lovingly but systematically engineered genius. In 1911, an article in a national magazine reported these plans: [1]

> Professor Leo Wiener of Harvard University . . . believes that the secret of precocious mental development lies in early training. . . . He is the father of four children, ranging in age from four to sixteen; and he has had the courage of his convictions in making them the subject of an educational experiment. The results have . . . been astounding, more especially in the case of his oldest son, Norbert.
>
> This lad, at eleven, entered Tufts College, from which he graduated in 1909, when only fourteen years old. He then entered the Harvard Graduate School.

Norbert completed his examinations and his doctoral dissertation in mathematical logic when he was eighteen, then studied with Bertrand Russell in Cambridge and David Hilbert in Göttingen, where he later crossed paths with von Neumann, nine years his junior, also a student of Hilbert's, and a world-renowned authority in several of Wiener's fields of interest. One of the most immediate differences between the two prodigies, even this early in their careers, was the pronounced contrast between their personalities.

Rare was the teacher or student who failed to be charmed by von Neumann, who went out of his way to assure fellow humans that he was just as mortal as everyone else. Wiener, an insecure, far less worldly, sometimes vain, and often hypersensitive personality, simply didn't go to as much trouble to make an impression outside the realm of mathematics, where he was confident to the point of arrogance. Bertrand Russell wrote of Wiener, in a letter to a friend: [2]

> At the end of Sept. an infant prodigy named Wiener, Ph.D. (Harvard), aged 18, turned up with his father who teaches Slavonic languages there, having first come to America to found a vegetarian communist colony, and having abandoned that intention for farming,

and farming for the teaching of various subjects. . . . The youth has been flattered, and thinks himself God Almighty—there is a perpetual contest between him and me as to which is to do the teaching.

Like Babbage, Wiener was famous for the feuds he carried on. While a student at Göttingen, he impressed the administrative head of the university, Richard Courant, but Wiener accused him of misappropriating several of the younger man's mathematical ideas and appending Courant's own name to them. When he returned to Cambridge, the outraged young genius turned his energies to a novel that was never published, about somebody who bore a remarkable resemblance to Courant, and who was depicted as a man who stole the ideas of young geniuses.

Before World War I, Wiener wrote pieces for the *Encyclopedia Americana*, taught philosophy at Harvard and mathematics at the University of Maine. During World War I, Private Wiener was assigned to the U.S. Army's Aberdeen Proving Grounds in Maryland, where he was one of the mathematicians responsible for the computation of firing tables. His service in 1918 was one of the reasons it was natural for Wiener's friend Vannevar Bush to think of Norbert thirty years later, when the Allies needed a way to put firing tables directly into the radar-guided aiming mechanism of antiaircraft guns.

After the end of World War I, Norbert Wiener joined the Massachusetts Institute of Technology as an instructor of mathematics. It turned out to be the beginning of his lifelong association with that institution. By the early 1920s, like his fellow polymath across the Atlantic, Wiener was turning out world-class papers in mathematics, logic, and theoretical physics. At MIT Wiener began his long friendship with Vannevar Bush, a man who in the early 1930s was deeply involved in the problems of building mechanical calculators, and in the 1940s took charge of the largest-scale administration of applied science in history.

Decades later, Wiener quarreled with his lifelong friend because Bush didn't side strongly enough with Wiener in his feud with two other colleagues. Such feuds were one of the more well-known characteristics of Wiener's style—he tended to take disagreements over scientific issues as personal attacks, even if the disputes involved his closest friends. Like Babbage, his judgment did not always seem equal to his imagination.

It must be said that Wiener did have many warm lifelong friendships that didn't go sour. For all his moodiness and paranoia, Wiener truly cared about "the human use of human beings" (as he was to title one of his later books on the implications of cybernetics), and passionately reminded the scientific community of their special responsibilities regarding the apocalyptic weaponry they had created. Despite his failure to get along with some of his colleagues, Wiener never wavered in his belief that the future of scientific enterprise lay in interdisciplinary cooperation. His friendship with the physiologist Arturo Rosenblueth, and their shared dream of stimulating such interdisciplinary pursuits, catalyzed the origins of cybernetics. But Wiener might never have worked with Rosenblueth if it wasn't for the Battle of Britain.

Like von Neumann, Wiener's most important need was for interesting problems. Like von Neumann, he knew that the quantum revolution was the most interesting problem of the 1920s. And one of the effects of quantum physics on the young mathematician's thinking was to convince him that some of the most interesting problems of purely theoretical mathematics could end up having the most concrete applications to the real world.

Another effect of quantum physics was the emergence of the importance of *probability* and statistical measures for dealing with phenomena based on uncertain information. Wiener's familiarity with these concepts was to mature under unexpected circumstances. Like von Neumann and Goldstine and Mauchly and Eckert, in the late 1930s Wiener wasn't yet aware that ballistics would be the avenue for bringing his knowledge of probability and statistics to bear on the most pragmatic problems, eventually to yield the most astonishing results. But, like them, he would soon come to understand that his war-related task was leading to profound scientific consequences far beyond the bounds of ballistics.

The scene was set for the emergence of Wiener's astonishing results, not by any series of scientific events, but by the political circumstances of the early 1940s. When war broke out in Europe, Bush assigned Wiener to the antiaircraft control project at MIT, under the direction of Warren Weaver, himself a distinguished mathematician. It seemed like a natural step for Wiener, considering his prior experience in the early ballistic calculation efforts at Aberdeen during World War I.

The key ideas that led to computers were in the air in the late 1930s, albeit in the rather rarefied air of metamathematics and other esoteric intellectual disciplines. The necessities of war and the coordinated scientific effort that they entailed served to bring those key ideas together with the few people who were equipped to understand them rather more quickly and urgently than might have happened in more normal times.

Von Neumann and Goldstine's accidental meeting at Aberdeen was fortuitous and unlikely, but could hardly be called incredible. One of the circumstances that brought Wiener together with the problem of antiaircraft guns, however, was downright weird. The technological turning point of the Battle of Britain, and a critical chapter in the science of communication systems in machines and organisms, originated when a young Bell Laboratories employee in America had an odd dream. The crucial dream was not about mathematics or engineering problems connected with computers, but was related to technical issues involving antiaircraft artillery. And it was the question of how to deal with dive bombers that was the rather urgent if indirect problem that led to Wiener's later insights.

The pathway between military strategy and scientific theory was far too circuitous, coincidental, and unlikely to have been predicted in advance, and became clearly discernible only in retrospect. In many respects, the birth of cybernetics was the kind of story more likely to be found in a novel than a scientific journal. One of the historical coincidences was the position of Vannevar Bush as the leader of U.S. war-related research. In his role as a research administrator, Bush knew that antiaircraft technology was one of his top priorities. As a scientist, MIT researcher, and friend of Norbert Wiener's, Bush was also concerned with the task of building high-speed mechanical calculators.

The Allies' two most pressing problems in the early years of World War II were the devastating U-boat war in the North Atlantic and the equally devastating Luftwäffe attacks on Britain. Turing's secret solution to the naval Enigma machine was responsible, in large part, for solving the U-boat problem. But where Turing's problem was one of cryptanalysis, of mathematically retrieving the meaning from a garbled message, the Luftwäffe problem was one of predicting the future: How can you shoot at a plane that is going as fast as your bullets?

Radar made it possible to track the positions of enemy aircraft, but there was no way to translate the radar-provided information into a ballistic equation quickly enough to do any good. And attacking airplanes had a disconcerting habit of taking evasive action. Vannevar Bush was well acquainted with the calculation problem when Bell Laboratories came to him with an interesting idea for an electrically operated aiming device. That is where the young engineer's dream came in.

His name was D. B. Parkinson, and he was working with a group of Bell engineers on an automatic level recorder for making more accurate measurements of telephone transmissions—a "control potentiometer," they called it. In the spring of 1940, Parkinson had the following dream: [3]

> I found myself in a gun pit or revetment with an anti-aircraft gun
> crew. . . . There was a gun there which looked to me—I had never

had any close association with anti-aircraft guns, but possessed some general information on artillery—like a 3 inch. It was firing occasionally, and the impressive thing was that *every shot brought down an airplane!* After three or four shots one of the men in the crew smiled at me and beckoned me to come closer to the gun. When I drew near he pointed to the exposed end of the left trunnion. Mounted there was the control potentiometer of my level recorder! There was no mistaking it—it was the identical item.

The electrical device, as it happened, was a good start on an automatic aiming mechanism. But very serious theoretical and mathematical problems, having to do with the way the control device sent and received instructions, cropped up when they tried to construct such a mechanism. That is when Bush turned to Weaver and Wiener.

During this wartime mathematical work related to radar-directed anti-aircraft fire, Wiener recognized the fundamental relationship between two basic problems—communication and control. The communication problem in the earliest days of radar was that the radar apparatus was like a badly tuned radio receiver. The true signal of attacking planes was often drowned out by false signals—noise—from other sources. Wiener recognized that this too was a kind of cryptography problem, if the true location of the enemy aircraft is seen as a message that must somehow be decoded from the surrounding noise.

The noisy radar was more than an ordinary "interesting problem," because once you understand messages and noise in terms of order and information measured against disorder and uncertainty, and apply statistics to predict future messages on the basis of information about past messages, it becomes clear (to a mathematician of Wiener's stature) that the issue is related to the basic processes of order and disorder in the universe. Once it is seen in statistical and mathematical terms, the communication problem leads to the heart of something more important, called information theory. But that branch of the story belongs to Claude Shannon as much as, or more than, it does to Wiener.

The control problem was where Wiener, and his very young and appropriately brilliant assistant, an engineer by the name of Julian Bigelow, happened upon the general importance of feedback loops. Assuming that it is possible to feed information about a plane's path into the aiming apparatus of a gun, how can that information be used to predict the probable location of the plane? The use of statistics and probability theory was one clue. A method for predicting the end of a message based on information about the beginning was another clue. The device in Parkinson's dream was another clue.

Then it occurred to Wiener and Bigelow that the human organism had already solved the problem they were facing. How is any human being, or chimpanzee for that matter, able to reach out a hand and pick up a pencil? How are people able to put one foot in front of the other, fall face-forward for a short distance, and end up taking a step? Both processes involve continuous, precise readjustments of muscles (the servomechanisms that move the gun), guided by continuous visual information (radar), controlled by a continuous process of predicting trajectories. The prediction and control take place in the nervous system (the control circuits of the aiming automata).

Wiener and Bigelow looked more closely at other servomechanisms, including self-steering machines as simple as thermostats, and concluded that *feedback* is the concept that connects the way brains, automatic artillery, steam engines, autopilots, and thermostats perform their functions. In each of those systems, some small part of the past output is fed back to the central processor as present input, in order to steer future output. Information about the distance of the hand from the pencil, as seen by the eye, is fed back to the muscles controlling the hand. Similarly, the position of the gun and the position of the target as sensed by radar are fed back to the automatic aiming device.

The MIT team wondered whether someone more informed about neurophysiology had come across analogous mathematics of pencil pushing, with similar results. As it happened, there was another team that, like Wiener and Bigelow, was made up of one infant prodigy and one slightly older genius, by the names Pitts and McCulloch respectively, who were coming down exactly the same trail from the other direction. A convergence of ideas that was both forced and fortuitous, related to but distinctly different from the convergence on digital computation, was taking place under the pressure of the war.

Even von Neumann was due to get into the act, as Wiener wanted him to do—Wiener persuaded MIT to try to outbid Princeton for von Neumann's attentions after the war. Politically, militarily, and scientifically, Wiener's corner of the plot was getting thick. The antiaircraft problem, the possible explanations for how brain cells work, the construction of digital computers, the decoding of messages from noise—all these seemingly unrelated problems were woven together when the leading characters were brought together by the war.

The founding of the interdisciplinary study that was later named *cybernetics* came about when Wiener and Bigelow wondered whether any process in the human body corresponded to the problems of excessive feedback in servomechanisms. They appealed to an authority on physiology, from the Instituto Nacional de Cardología in Mexico City. Dr. Arturo Rosenblueth replied that there was exactly such a pathological condition named (meaningfully) the *purpose tremor*, associated with injuries to the cerebellum (a part of the brain involved with balance and muscular coordination).

Together the mathematician, the neurophysiologist, and the engineer plotted out a new model of nervous system processes that they believed would demonstrate how purpose is embodied in mechanism—whether that mechanism is made of metal or flesh. Wiener, never reluctant to trumpet his own victories, later noted that this conception "considerably transcended that current among neurophysiologists." [4]

Wiener, Bigelow, and Rosenblueth's model, although indirectly derived from top-secret war work, had such general and far-reaching implications that it was published under the title "Behavior, Purpose and Teleology," in 1943, in the normally staid journal *Philosophy of Science*.[5] The model was first discussed for a small audience of specialists, however, at a private meeting held in New York in 1942, under the auspices of the Josiah Macy Foundation. At that meeting was Warren McCulloch, a neurophysiologist who had been corresponding with them about the mathematical characteristics of nerve networks.

McCulloch, a neurophysiologist based at the University of Illinois, was, naturally enough in this company, an abnormally gifted and colorful person who had a firm background in mathematics. One story that McCulloch told about himself goes back to his student days at Haverford College, a Quaker institution. A teacher asked him what he wanted to do with his obviously brilliant future: [6]

> "Warren," said he, "what is thee going to be?" And I said, "I don't know," "And what is thee going to do?" And again I said, "I have no idea, but there is one question I would like to answer: What is a number, that man may know it, and a man that he may know a number?" He smiled and said, "Friend, thee will be busy as long as thee lives."

Accordingly, the mathematician in McCulloch strongly desired a tool for reducing the fuzzy observations and theoretical uncertainties of neurophysiology to the clean-cut precision of mathematics. Turing, and Bertrand Russell before him, and Boole before that, had been after something roughly similar, but they all lacked a deep understanding of brain physiology. McCulloch's goal was to find a basic functional unit of the brain, consisting of some combination of nerve cells, and to discover how that

Cyberneticist Warren McCulloch, circa 1967. (Courtesy of the MIT Museum.)

basic unit was built into a system of greater complexity. He had been experimenting with models of "nerve networks" and had discovered that these networks had certain mathematical and logical properties.

McCulloch started to work with a young logician by the name of Walter Pitts. Pamela McCorduck, historian of artificial intelligence research, attributes to Manuel Blum, a student of McCulloch's and now a professor at the University of California, the story of Pitts' arrival on the cybernetic scene. At the age of fifteen, Walter Pitts ran away from home when his father wanted him to quit school and get a job. He arrived in Chicago, and met a man in a park who knew a little about logic. This man, "Bert" by name, suggested that Pitts read a book by the logician Carnap, who was then teaching in Chicago. Bert turned out to be Bertrand Russell, and Pitts introduced himself to Carnap in order to point out a mistake the great logician had made in his book.

Pitts studied with Carnap, and eventually came into contact with McCulloch, who was interested in consulting with logicians in regard to his neurophysiological research. Pitts helped McCulloch understand how certain kinds of networks—the kinds of circuits that might be important parts of nervous systems as well as electrical devices—could embody the logical devices known as Turing machines.

McCulloch and Pitts developed a theory that regarded nerves as all-or-none, on-or-off, switchlike devices, and treated the networks as circuits that could be described mathematically and logically. Their paper, "A Logical Calculus of the Ideas Immanent in Nervous Activity," was published in 1943 when Pitts was still only eighteen years old.[7] They felt that they were only beginning a line of work that would eventually address the questions of how brain physiology is linked to knowledge.

When Wiener, Bigelow, and Rosenblueth got together with McCulloch and Pitts, in 1943 and 1944, a critical mass of ideas was reached. Pitts joined Wiener at MIT, then worked with von Neumann at the Institute for Advanced Study after the war. By the time this interdisciplinary cross-fertilization was beginning, the ENIAC project had progressed far enough for digital computers to join the grand conjunction of ideas.

A series of meetings occurred in 1944, involving an interdisciplinary blend of topics that seemed to be coming together from subject areas as far afield as logic, statistics, communication engineering, and neurophysiology. The participants were an equally eclectic assortment of thinkers. It was at one of these meetings that von Neumann made the acquaintance of Goldstine, whom he was to encounter again not long afterward, at the Aberdeen railroad station. Rosenblueth had to depart for Mexico City in 1944, but by December, Wiener, Bigelow, von Neumann, Howard Aiken of the Harvard-Navy-IBM Mark I calculator project, Goldstine, McCulloch, and Pitts formed an association they called "The Teleological Society," for the purpose of discussing "communication engineering, the engineering of control devices, the mathematics of time series in statistics, and the communication and control aspects of the nervous system."[8] In a word—cybernetics.

In 1945 and 1946, at the Teleological Society meetings, and in personal correspondence, Wiener and von Neumann argued about the advisability of placing too much trust in neurophysiology. Von Neumann thought that the kind of tools available to McCulloch and Pitts put brain physiologists in the metaphorical position of trying to decipher computer circuits by bashing computers together and studying the wreckage.

To von Neumann, the bacteriophage—a nonliving microorganism that can reproduce itself—was a much more promising object of study. He felt that much more could be learned about nature's codes by looking at microorganisms than by studying brains. The connection between the mysteries of brain physiology and the secrets of biological reproduction were later to emerge more clearly from theories involving the nature of information, and von Neumann turned out to be right—biologists were to make faster progress in understanding the coding of biological reproduction than neuroscientists were to make in their quest to decode the brain's functions.

The Macy Foundation, which had sponsored the meetings that led to the creation of the Teleological Society, continued to sponsor free-wheeling meetings. Von Neumann and Wiener were the dramatic costars of the meetings, and the differences in their personal style became part of the excited and dramatic debates that characterized the formative years of cybernetics. Biographer Steve Heims, in his book about the two men—*John von Neumann and Norbert Wiener*—noted the way their contrasting personae emerged at these events: [9]

Wiener and von Neumann cut rather different figures at the semiannual conferences on machine-organism parallels, and each had his own circle of admirers. Von Neumann was small and plump, with a large forehead and a smooth, oval face. He spoke beautiful and lucid English, with a slight middle-European accent, and he was always carefully dressed; usually a vest, coat buttoned, handkerchief in pocket, more the banker than the scholar. He was seen as urbane, cosmopolitan, witty, low-key, friendly, and accessible. He talked rapidly, and many at the Macy meetings often could not follow his careful, precise, rapid reasoning. . . .

Wiener was the dominant figure at the conference series, in his role as brilliant originator of ideas and enfant terrible. Without his scientific ideas and his enthusiasm for them, the conference series would never have come into existence, nor would it have had the momentum to continue for seven years without him. A short, stout man with a paunch, usually standing splay-footed, he had coarse features and a small white goatee. He wore thick glasses and his stubby fingers usually held a fat cigar. He was robust, not the stereotype of the frail and sickly child prodigy. Wiener evidently enjoyed the meetings and his central role in them: sometimes he got up from his chair and in his ducklike fashion walked around and around the circle of tables, holding forth exuberantly, cigar in hand, apparently unstoppable. He could be quite unaware of other people, but he communicated his thoughts effectively and struck up friendships with a number of the participants. Some were intrigued as much as annoyed by Wiener's tendency to go to sleep and even snore during a discussion, but apparently hearing and digesting what was being said. Immediately upon waking he often would make penetrating comments.

Although the nerve network theory was to suffer a less than glorious fate when neurophysiology progressed beyond what was known about nerve cells in the 1940s, the nerve-net models had already profoundly influenced the design of computers. (Later research showed that switching circuits are not such an accurate model for the human nervous system, because neurons do not act strictly as "all-or-none" devices.) Despite his misgivings about the state of the art in theories of brain functioning, in his 1945 "First Draft," von Neumann adopted the logical formalism proposed by McCulloch and Pitts. When the architectural template of all future general-purpose computers was first laid down, the cyberneticists' findings influenced the logical design.

In 1944 and 1945, Wiener was already thinking about a scientific model involving communication, information, self-control—an all-embracing way of looking at nature that would include explanations for computers and brains, biology and electronics, logic and purpose. He later wrote: "It became clear to me almost at the very beginning that these new concepts of communication and control involved a new interpretation of man, of man's knowledge of the universe, and of society. "[10]

Wiener was convinced that biology, even sociology and anthropology, were to be as profoundly affected by cybernetics as electronics theory or computer engineering; in fact, anthropologist Gregory Bateson was closely involved with Wiener and later with the first AI researchers. While Shannon established information theory, and von Neumann pushed the development of computer technology, Wiener retreated from the politics of big science in the postwar world to articulate his grand framework.

After the war, as the plans for the Institute for Advanced Study's computer proposed by von Neumann were put into action, with Julian Bigelow as von Neumann's chief engineer on the project, and as Mauchly and Eckert struck out on their own to start the commercial computer industry, Wiener headed for Mexico City to work with Rosenblueth. Then, in the spring of 1947, Wiener went to England, where he visited the British computer-building projects, and spoke with Alan Turing.

When he returned to Mexico City, Wiener wrote his book and decided to title it and the new field *Cybernetics*, from the Greek word meaning "steersman." It was subtitled: *or Control and Communication in the*

Norbert Wiener at MIT, with his ever-present cigar. (Courtesy of the MIT Museum.)

Animal and the Machine. Cybernetics was the description of a general science of mechanisms for maintaining order in a disorderly universe, the process of steering a course through the random forces of the physical world, based on information about the past and forecasts about the future.

When a steersman moves a rudder, the craft changes course. When the steersman detects that the previous change of course has oversteered, the rudder is moved again, in the opposite direction. The feedback of the steersman's senses is the controlling element that keeps the craft on course. Wiener intended to embed in the name of the discipline the idea that there is a connection between steering and communication. "The theory of control in engineering, whether human or animal or mechanical," he stated, "is a chapter in the theory of messages." [11]

The mathematics underlying the steering of rudders or antiaircraft guns and the steering of biological systems was the same—it was a general law, Wiener felt, like the laws of motion or gravity. Wiener's intuitions turned out to be correct. Communication and control, coding and decoding, steering and predicting, were becoming more important to physicists and biologists, who were interested in phenomena very different from guns or computing machines.

In the late 1940s, another new category of interdisciplinary theorists who would come to be known as molecular biologists were beginning to think about the *coding* mechanism of genetics. Even the quantum physicists were looking into the issues that were so dear to Wiener, Bigelow, and Rosenblueth. It looked as if Wiener might be onto an even more cosmic link between information, energy, and matter. A scientific watershed was imminent, and many of his colleagues were expecting more major breakthroughs from Wiener. By the fall of 1947, prior to its 1948 publication, his book on cybernetics was making the rounds of government and academic experts in manuscript form.

Robert Fano, a professor of electrical engineering who eventually became head of the electrical engineering department at MIT and administrative leader of MIT's pioneering computer project known as MAC, witnessed some strange behavior on Wiener's part around that time, behavior that Fano later had cause to remember when Claude Shannon published his work. Fano was working on his doctoral thesis in electrical engineering. From time to time, Wiener would walk into the student's office, inform him rather cryptically that "information is entropy," and walk out without saying another word.[12]

By the end of 1946, Wiener had reached a decision that had nothing to do with the cold formalisms of mathematics, a decision that distinguished him in yet another way from his weaponry-oriented colleague. Renouncing any future role in weapons-related research, Wiener deliberately removed himself from the hot center of the action in the development of computer technology (as opposed to cybernetic theory) when he stated: "I do not expect to publish any future work of mine which may do damage in the hands of irresponsible militarists." [13] Fortunately for Wiener, and for the scientific world, the implications of his discoveries were not limited to military applications. It quickly became evident that weapons were not the only things of interest that were built from communication and control codes.

By the late forties and early fifties, the atmosphere was crackling with new scientific ideas having to do with what nobody yet called information theory. The quantum physicist Erwin Schrodinger gave a famous lecture at Cambridge University in 1945, later published, on the topic "What Is Life?" One of the younger physicists in the audience, Francis Crick, decided to switch to biology, where the most crucial decoding in scientific history was waiting for him. Von Neumann turned out to be right in his dispute with Wiener—the bacteriophage, not the nervous system, was the subject of the next great decoding.

Von Neumann's ideas about self-reproducing automata—patterns complex enough and highly ordered enough to direct their own replication—seemed to point toward the same idea. Something about order and disorder, messages and noise, was near the heart of life. The manipulation of information looked like something more than a game mathematicians play, even more than a new capability of machines. Information, in a way that was not mathematically demonstrated until Claude Shannon's 1948 publications, began to look like a reflection of the way the universe works. The whole idea was a major wrenching of mind-set, first for scientists, then for many others.

At the beginning of the twentieth century, scientists saw the universe in terms of particles and forces interacting in complicated but orderly patterns that were, in principle, totally predictable. In important ways, all of the nonscientists who lived in an increasingly mechanized civilization also saw the universe in terms of particles and forces and clockwork cosmos. Around sixty years ago, quantum theory did away with the clockwork and predictability. Around thirty years ago, a few people began to look at the world and see, as Norbert Wiener put it, "a myriad of To Whom It May Concern messages."

The idea that information is a fundamental characteristic of the cosmos, like matter and energy, is still young, and further surprise discoveries and applications are sure to pop up before a better model comes along.

Before the 1950s, only scientists thought about the idea that information had anything to with anything. Common words like *communication* and *message* were given new, technical meanings by Wiener and Claude Shannon, who independently and roughly simultaneously demonstrated that everything from the random motions of subatomic particles to the behavior of electrical switching networks and the intelligibility of human speech is related in a way that can be expressed through certain basic mathematical equations.

The information-related equations were useful in building computers and telephone networks, but they also had significant impact on all the sciences. Research inspired by the information-communication model has provided clues to some of the fundamental features of the universe, from the way the cellular instructions for life are woven into the arrangement of atoms in DNA molecules, to the process by which brain cells encode memory. The model has become what Thomas Kuhn calls a "scientific paradigm." The two fundamental pillars of this paradigm were Claude Shannon's information theory and Wiener's cybernetics.

The significance of these two theoretical frameworks that came to the attention of scientists in the late 1940s and began to surface in public consciousness in the 1950s, and the mass attitude-shift they implied, was noted by Pamela McCorduck, in her history of artificial intelligence research: [14]

> *Cybernetics* recorded the switch from one dominant model, or set of explanations for phenomena, to another. *Energy*—the notion central to Newtonian mechanics—was now replaced by *information*. The ideas of information theory, such as coding, storage, noise, and so on, provided a better explanation for a whole host of events, from the behavior of electronic circuits to the behavior of a replicating cell. . . . These terms mean pretty nearly what you'd think. *Coding* refers to "a system of signals used to represent letters or numbers in transmitting messages"; *storing* means holding these signals until they're needed. *Noise* is a disturbance that obscures or affects the quality of a signal (or message) during transmission.

It turns out that coding and storing happen to be central problems in the logical design of computing machines and the creation of software. The basic scientific work that resulted in information theory did not originate from any investigation of computation, however, but from an analysis of communication. Claude Shannon, several years younger than Turing, working about a year after the British logician's discoveries in metamathematics, did another nifty little bit of graduate work that tied together theory and engineering, philosophy, and machinery.

CHAPTER SIX

Inside Information

His unicycle skills notwithstanding, Claude Shannon has been less flamboyant but no less brilliant than his elder colleagues. Rather than advertising his own genius like Wiener, or blitzing the world of science with salvo after salvo of landmark findings like von Neumann, Claude Shannon has published unprolifically, and he spends more time attempting to diminish rather than embellish the mythology that grew up around his infrequent but monumental contributions. A modest man, perhaps, but hardly a timid one: When Shannon has something to publish, it usually changes the world.

Claude Shannon was a bona fide prodigy, twenty-two years old when he published (in 1937) the famous MIT master's thesis that linked electrical circuitry to logical formalisms. He was the peer of pioneers like Turing, Wiener, and von Neumann, the teacher of the first generation of artificial intelligence explorers like John McCarthy and Marvin Minsky, and the mentor of Ivan Sutherland, who has been one of the most important contemporary infonaut-architects.

When Shannon's papers establishing information theory were published in 1948, he was thirty-two. The impact on science of this man's career was incalculable for these two contributions alone, but he also wrote a pioneering article on the artificial intelligence question of game-playing machines, published in 1950. In 1953, at about the same time von Neumann and Turing were both thinking about the mathematical possibilities of self-reproducing machinery, Shannon published another major work on the subject of these special automata.

In 1956, at the age of forty, Shannon was one of the organizers of the conference at Dartmouth that gave birth to the field of artificial intelligence. From the prewar discoveries that scooped Wiener and von Neumann, to the explorations in the 1950s that led to both AI and multiaccess computer systems, his life and ideas formed the single most important bridge between the wartime origins of cybernetics and digital computers and the present age of artificial intelligence and personal computing.

What Shannon did in 1937 was to provide a way to design machines based on the logical algebra described a century before by George Boole. Boole, in *The Laws of Thought*, stated that he had succeeded in connecting the processes of human reason to the precise symbolic power of mathematics. There were only two values in the logical calculation system that Boole proposed: 1 and 0. If a statement is true, it can be designated by the symbol 1; and if it is false, the symbol 0 can be used. In this system, a *truth table* describes the various possible logical states of a system. Given an input state, a truth table for a specific operation determines the appropriate output state whenever that operation is applied to that input. Another way of saying that would be that given a starting tape, the truth table determines what the ending tape will be.

In Boolean algebra, one fundamental logical operation is *not*, an operation that reverses the input, so that the output of a "not" operation is the opposite of the input (remember there are only two symbols or states). Another fundamental operation is *and*, which dictates that the output is true (or "on" or "1") if and only if every one of several inputs are also true ("on," "1"). For example, the listing in the table for "A is true and B is true" would be set for "1" when A is "1" and B is "1," and set for "0" in all other cases. One could look up the answer in the truth table by finding the input row where both A and B are equal to 1:

NOT

AND

Input	Output		Input A	Input B	Output
0	1		0	0	0
1	0		0	1	0
			1	0	0
			1	1	1

The way that results are determined by matching the proper rows and columns in the truth tables, a purely automatic procedure, has a crucial resemblance to the "instruction tables" Turing proposed.

One of the important features of Boolean algebra is the way logical operations can be put together to form new ones, and collections of logical operations can be put together to perform arithmetic operations. Logical syllogisms can be constructed in terms of operations on zeroes and ones, by arranging for the output of one truth table to feed input to another truth table. For example, it turns out that by putting a "not" before every "and" input, and putting another "not" after its output, it is possible to build an "or" operation. By stringing together various sequences of only these two basic operations, "not" and "and," it is possible to build procedures for adding, subtracting, multiplying, and dividing. Logic and arithmetic are thus intimately and simply related. What nobody knew until Shannon told us was that the same algebra could describe the behavior of electrically switched circuits.

Equally important was the way these combinations of logical and arithmetic operations could be used to build a "memory" operation. Boolean algebra makes it possible to devise a procedure, or build a device, the "state" of which can store specific information—either data or operations. If electrical circuitry can perform logical and mathematical operations, and can also store the results of those operations, then electronic digital computers can be designed.

Until Shannon, Boolean algebra had been a curious and almost totally forgotten eddy in the mainstream of mathematical thought for almost a century, and was certainly unknown to the more practical-minded world of physics and electrical engineering. And that is where the genius of Shannon's rediscovery lies, for he was writing a thesis in electrical engineering, not mathematical logic, and the objects of his concern were not the processes of thought but the behavior of large numbers of electrical switches connected together into the kinds of circuits one finds in a telephone system.

Shannon was interested in the properties of complicated electrical systems that were built from very simple devices known as *relays*. A relay is a switch—a device that opens or closes a circuit, permitting or blocking the flow of electrical current—not unlike an ordinary light switch, except a relay is not switched on or off by a human hand, but by the passage of an electrical current.

A relay contains an electromagnet. When a small current flows into the relay, the electromagnet is activated, closing the circuit controlled by the relay until the input current is turned off. In other words, the electromagnet is a small electrical circuit that opens and closes another electrical circuit. The circuit of one relay can also control the electromagnet of the next relay, and so on, until you have a complete circuit that is made of nothing but switches, all controlling one another, depending on how they are set at the beginning and how they are altered by new input.

Each relay and the circuit controlled by that relay can be in only one of two states, on or off. This two-state characteristic of switched circuits is what links electricity to logic, for each relay-controlled circuit can be seen as a truth table, where current flows from the output only when specified input conditions are satisfied, and logical operations can be seen as physical devices that emit an output pulse if and only if all of their input switches are on, or off, or some specified combination.

In the 1930s, telephone systems were using ever larger and more complicated mazes of circuits controlled by these relays. Instead of requiring a human operator to plug the proper jack into the right part of a switchboard, relays could close the circuit when the specified input conditions were reached. Using relays, all kinds of useful things could be done in the way of automatic dialing and routing. But the growing complexity of the circuitry was getting to be a problem. It was becoming harder and harder to figure out what these big collections of switches were doing.

Shannon was looking for a mathematical procedure that was best suited for describing the behavior of relay circuits. His thesis showed how George Boole's algebra could be used to describe the operations of these complex circuits. And he was not unaware of the implications of the fact that circuits could now be designed to represent the operations of logic and arithmetic.[1]

If logic was the formal system that most closely matched the operations of human reason, and if Boole's truth tables could embody such a formal system of simulated reasoning, then with the use of truth tables as the "instruction tables" Turing discussed, and with switching devices like relays to represent the "states" of the machine (or the cells of the tape), it would be possible to build electrical circuits that could simulate some of the logical operations of human thought.

When the digital computer builders got together to plan the future development of the technology, Shannon was in the thick of it—and he didn't hesitate to remind his colleagues that what they were building was the first step toward an artificial intelligence. But during the ten years immediately following his first breakthrough, Shannon turned to a different aspect of this new field. His employer was Bell Laboratories, and the electrical or electronic communication of messages was his specialty. AT&T, the foremost communication company in the world, was the owner of Bell Laboratories, so naturally the laboratory was interested in supporting Shannon's probes into the fundamental nature of communication. Shannon was encouraged to pursue his interest in questions such as: When something is communicated, what is delivered from one party to another? When a communication is obscured by noise or encryption, what fails to get across?

This was the communication part of the communication and control problem pointed out by Wiener. During the war, working at top-secret defense projects for Bell Laboratories, Shannon was involved in cryptological work that brought him into contact with Turing. After the war, Shannon concentrated on describing the nature of the entity they were communicating and manipulating with all these logical and mathematical circuits.

At this point, nobody knew, exactly, what information was. Just as he had found the perfect tool for describing relay circuits, after the war Shannon wanted to find mathematical tools for precisely defining the invisible but powerful commodity that these new machines were processing. He succeeded in finding the descriptive tools he sought, not in an obscure corner of mathematics, as in the case of Boole's algebra, but in the fundamental laws governing energy.

Like Turing, Shannon put the surprise finishing touch on a project that scientists had worked at for centuries. In this case, the quest was not to understand the nature of symbol systems, but a more pragmatic concern with the nature of energy and its relation to information. Although Shannon was specifically looking at the laws underlying the communication of messages in man-made systems, and generally interested in the difference between messages and noise, he ended up dealing with the laws governing the flow of energy in the universe. In particular, he discovered the secrets of decoding telephone switching networks, hidden in the work of previous scientists who had discovered certain laws governing heat energy in steam engines.

Back when the Industrial Revolution was getting started, and steam-powered engines were the rage, it became a practical necessity to find out something about the efficiency of these energy-converting devices. In the process, it was discovered that something fundamental to the nature of heat prevents any machine from ever being perfectly efficient. The study of the movement of heat in steam engines became the science of thermodynamics, given precise expression in 1850 by Rudolf Clausius, in his two laws of thermodynamics.

The first law of thermodynamics stated that the energy in closed systems is constant. That means that energy can neither be created nor destroyed in such systems, but can only be transformed. The second law states, in effect, that part of that unchangeable reservoir of energy becomes a little less usable every time a transformation takes place. When you pour hot water into cold water, you can't separate it back into a hot and a cold glass of water again (without using a lot more energy). *Entropy*, meaning "transformation," was the word Clausius later proposed for that lost quantity of usable energy.

Entropy as defined by Clausius is not just something that happens to steam engines or to glasses of water. It is a universal tendency that is as true for the energy transactions of the stars in the sky as it is for the tea kettle on the stove. Because the universe is presumed to be a closed system, and since Clausius demonstrated that the entropy of all such systems tends to increase with the passage of time, the gloomy prediction of a distant but inevitable "heat death of the universe" was a disturbing implication of the second law of thermodynamics. "Heat death" was what they called it because heat is the most entropic form of energy.

But the gloomy news about the end of time wasn't the only implication of the entropy concept. When it was discovered that heat is a measure of the average motion of a population of molecules, the notion of entropy became linked to the measure of order or disorder in a system. If this linkage of such disparate ideas as "heat," "average motion," and "order of a system" sounds confusing, you have a good idea of how nineteenth-century physicists felt. For a long time, they thought that heat was some kind of invisible fluid that was transferred from one object to another. When it was discovered that heat is a way of characterizing a substance in which the molecules were, on the average, moving around faster than the molecules in a "cold" substance, a new way of looking at systems consisting of large numbers of parts (molecules, in this case) came into being. And this new way of looking at the way the parts of systems are arranged led, eventually, to the entropy-information connection.

Because "average motion" of molecules is a statistical measure, saying something about the amount of heat in a system says something about the way the parts of that system are arranged. Think about a container of gas. The system in this case includes everything inside the container and everything outside the container. The gas is considered to be hot if the average energy of the molecules inside the container is higher than the average energy of the molecules outside the container. Some of the molecules inside the container might, in fact, be less energetic (colder) than some of the molecules outside the container—but on the average, the population of molecules inside are more energetic than the population of molecules outside.

There is a certain *order* to this arrangement—energetic molecules are more likely to be found inside the container, less energetic molecules are more likely to be found outside. If there were no container, the highly energetic molecules and the less energetic molecules would mix, and there would be no sharp differentiation between hot and cold parts of the system.

A system with high entropy has a low degree of order. A system with low entropy has a higher degree of order. In a steam engine, you have the heat in one place (the boiler) and it is dissipated into the cold part (the condenser). This is a very orderly (low-entropy) system in the sense that anyone can reliably predict in which part of the engine hot molecules are likely to be found. But when all the parts of a steam engine are the same temperature, and hot molecules are equally likely to be found in the boiler and the condenser (and hence the entropy is high), the engine can't do any work.

Another physicist, Boltzmann, showed that entropy is a function of the way the parts of the system *are arranged*, compared with the number of ways the system *can be arranged*. For the moment, let's forget about molecules and think about decks of cards. There is a large number of ways that fifty-two cards can be arranged. When they come from the factory, every deck of cards is arranged in a definite order, by suit and by value. With a little bit of thought, anybody can predict which card is the fifth from the top of deck. This predictability and orderliness disappears when the deck is shuffled.

An unshuffled deck of cards has a lower degree of entropy because energy went into arranging it in an unlikely manner. Less energy is then required to put the deck into a more probable, less orderly, less predictable, more highly entropic state: According to the second law of thermodynamics, all decks of cards in the universe will eventually be shuffled, just as all molecules will have an equal amount of energy.

James Clerk Maxwell, yet another nineteenth-century scientist, proposed a paradox concerning this elusive quality called entropy, which seems to relate such intuitively dissimilar measures as energy, information, order, and predictability. The paradox became infamous among physicists under the name "Maxwell's demon." Consider a container split by a barrier with an opening small enough to pass only one molecule at a time from one side to the other. On one side is a volume of hot gas, in which the average energy of the molecules is higher than the average energy of the molecules in the cold side of the container. According to the second law, the hotter, more active molecules should eventually migrate across to the other side of the container, losing energy in collisions with slower-moving molecules, until both sides reach the same temperature.

What would happen, Maxwell asked, if you could place a tiny imp at the molecular gate, a demon who didn't contribute energy to the system, but who could open and close the gate between the two sides of the container? Now what if the imp decides to only let the occasional slow-moving, colder molecule pass from the hot to the cold side when it randomly approaches the gate? Taken far enough, this policy would mean that the hot side would get hotter and the cold side would get colder, and entropy would decrease instead of increase, without any energy being added to the system!

In 1922, a Hungarian student of physics by the name of Leo Szilard (later to be von Neumann's colleague on the Manhattan Project), then in Berlin, finally solved the paradox of Maxwell's demon by demonstrating that the demon does indeed contribute energy to the system, but like a good magician, the demon does not expend that energy in its most visible activity—moving the gate—but in *what it knows* about the system. The demon is part of the system, and it has to do some work in order to differentiate the hot and cold molecules at the proper time to open the gate. Simply by obtaining the information about the molecules that it needs to know to operate the gate, the demon adds more entropy to the system than it subtracts.

Although Szilard showed implicitly that information and entropy were intimately connected, the explicit details of the relationship between these two qualities, expressed in the form of equations, and the generalization of that relationship to such diverse phenomena as electrical circuits and genetic codes, were not yet known. It was Claude Shannon who made information into a technical term, and that technical term has since changed the popular meaning of the word.

Another puzzle related to entropy, and the cryptic partial solution to it proposed in 1945 by another physicist, was a second clue linking it to information. Quite simply: If the universe tends toward entropy, how does life, a highly ordered, energy-consuming, antientropic phenomenon, continue to exist? In a universe flowing toward disorder, how on earth did one-celled creatures complicate themselves enough to build a human nervous sytem?

Quantum physicist Erwin Schrödinger pointed out that life on earth defies the cosmic energy tide courtesy of our sun. As long as the sun keeps shining, the earth is not a closed system. Photochemical reactions on earth capture a tiny fraction of the sun's radiant energy and use it to complicate things. In his famous "What Is Life?" lecture in 1945, Schrödinger remarked that "living organisms eat negative entropy." The relationship between negative entropy and information, like Boole's obscure algebra, was just waiting to be found when Shannon started to wonder how messages manage to maintain their order in a medium where disorder is often high.

The matter of devising a simple code and reliably transmitting it from place to place was very important to British cryptographers, and Shannon had done his own work in cryptography. The prediction of the behavior of the electrical circuits used to transmit messages made of these codes was another one of Shannon's interests. When he put it all together with a formal examination of how messages can be distinguished from noise, and found that the very equation he sought was a variation of the defining equation for entropy, Claude Shannon happened upon the fact that the universe plays twenty questions with itself.

The formal foundations of information theory were laid down in two papers in 1948, and at their core were fundamental equations that had a definite relationship to Boltzmann's equations relating entropy to the degree of order in a system. But the general idea behind the equations was simple enough for Shannon to suggest a game as a way of understanding the quantitative dimension of coding and communication.

The game is a mundane variation of "twenty questions." In the case of the English alphabet, it turns out to be a game of "five questions." Player number one thinks of a letter of the alphabet. Player number two tries to guess the letter, using only questions like "is it earlier than L in the alphabetical sequence?" It is a strictly yes-or-no game, in which only one of two possible answers applies at every move.

Shannon pointed out that it takes a maximum of five questions to locate any one of thirty symbols necessary for making English sentences. If the sequence of yes or no decisions needed to specify the correct letter is converted into a sequence of zeroes and ones or a sequence of on and off impulses, or any other kind of binary symbol, you have a code for communicating the alphabet—which is, in fact, the basis for the code used for transmitting teletypewriter messages.

This game can be visualized as a tree structure, where each letter is the only leaf on a branch that branches off a branch that eventually branches off a trunk. Or it can be seen as a garden of forking paths, where each path is a sequence of one-way-or-the-other decisions, and the location of any endpoint can be coded by specifying the sequence of decisions along the path. It is also a good way to locate an address in a computer's memory or to encode an instruction to be placed in that location. This basic element in this game-tree-code, the binary decision, was the basis for Shannon's basic measure of information—the *bit*. Whenever computer enthusiasts speak of a "bit," they are referring to one of those decisions in the garden of forking paths.

Note that each decision, each bit, reduces the uncertainty of the situation, whether you are designating turns in a pathway or numbers in a guessing game or the energy state of molecules in a container. But what if you were to use a different strategy to guess the right answer? What if you just named each one of the possible letters, one at a time, in sequence or randomly? What are the odds that any one of those guesses would be right? This relates to probability theory, the mathematical principles governing the random selection of small samples from large populations.

The relative probability of an event occurring, whether it is the probability of a molecule being hot or the probability of a symbol being a specific letter of the alphabet, depends on the total number of cases in the population and the frequency of the specified event. If there are only two cases in the population, a single yes or no decision reduces uncertainty to zero. In a group of four, it takes two decisions to be sure. In a group of trillions, you have to guess a little. When you are making predictions about such large populations, *averages* based on the overall behavior of the population have to replace precise case-by-case calculations based on the behavior of individual members of the population.

One of the properties of a statistical average is that it is quite possible for a population to be characterized by an average value that is not held by any particular element of the population. If you have a population consisting of three people, and you know that one is three feet tall, one is five feet tall, and one is six feet tall, you have quite precise information about that population, which would enable you to pick out individuals by height. But if all you know is that the average height of the population is four feet, eight inches, you wouldn't know anything useful about any one of the three particular individuals. Whenever a system is represented by an average, some information is necessarily lost, just as two energy states always lose a little energy when they are brought into equilibrium.

Whenever you move from an average measure to a precise measure, you have reduced your uncertainty about that population. And that reduction in uncertainty is where the statistical properties that govern the motions of populations of molecules are connected to the statistical properties of a binary code, where entropy meets information. To see how uncertainty can relate to a binary code, think about the game of twenty questions. If the object of the game is to guess a number between one and one hundred, and player one asks if the number is larger than fifty, an answer from player two (no matter whether it is yes or no) reduces player one's uncertainty by one half. Before asking the question, player one has one hundred possible choices. After asking that single yes or no question, player one either knows that the number is greater than fifty or that it is less than fifty.

One of the things Shannon demonstrated in 1948 was that the entropy of a system is represented by the logarithm of the number of possible combinations of states in that system—which is the same as the number of yes-or-no questions that have to be asked to locate one individual case. Entropy, as it was redefined by Shannon, is the same as the number of binary decisions necessary to identify a specific sequence of symbols. Taken together, those binary decisions, like the answers in the game, constitute a definite amount of information about the system.

When it comes to arranging molecules, living organisms seem to have a great deal of information about how to take elementary substances and turn them into complex compounds. Somehow, living cells manage to take the hodgepodge of molecules found in their environment and arrange them into the substances necessary for sustaining the life of the organism. From a disorderly environment, living creatures somehow create their own internal order. This remarkable property sounds suspiciously like Maxwell's demon. The answer, as we now know, is to be found in the way the DNA molecule arranges its elements—doing so in such a way that the processes necessary for metabolism and reproduction are encoded. The "negative entropy" that Schrödinger says is the nourishment of all life is information, and Shannon showed exactly how such coding can be done—in molecules, messages, or switching networks.

It has been said, by the way, that Shannon was reluctant to use the word "entropy" to represent this measure implied by his equations, but von Neumann told him to go ahead and use it anyway, because "since nobody knows what entropy is, in a debate you will be sure to have an advantage."

Remember that entropy is where Shannon ended up, not where he started. Hot molecules and DNA codes were far from his original intention. He got to the guessing game and the notion of bits and the relationship between uncertainty and entropy because he looked closely at what a message really is. How does a signal that conveys information differ from everything else that happens? How much energy must be put into broadcasting a voice over the radio to be sure that it will be understood despite atmospheric interference or static from other sources? These were the questions that Shannon set out to answer.

Shannon's 1948 publication ("A Mathematical Theory of Information") presented a set of theorems that were directly related to the economical and efficient transmission of messages on noisy media, and indirectly but still fundamentally related to the connection between energy and information.[2] Shannon's work was a direct answer to an engineering problem that had not decreased in importance since the war: how can messages be coded so that they will be reliably transmitted and received over a medium where a certain amount of noise is going to garble reception?

Shannon showed that any message can be transmitted with as high a reliability as one wishes, by devising the right code. The limit imposed by nature is concerned only with the limit of the communication channel. As long as there is a channel, no matter how noisy, a code can be devised to transmit any message with any degree of certainty. Entropy is a measure of the relationship between the complexity of the code and the degree of certainty. These theorems meant a lot to radio and telephone engineers, and made color television as well as broadcasts from the moon possible, but Shannon stated them in a way that demonstrated their universality beyond the domain of electrical engineering.

The key to life itself, in fact, turned out to be a matter of information, as the world learned five years later, when that young physicist-turned-biologist who had attended Schrödinger's lecture, Francis Crick, teamed up with James Watson to decipher the molecular genetic coding of the DNA helix. Scientifically, and on the level of public consciousness, people seemed to jump rather too quickly to make the transition from an energy-based metaphor of the universe to an information model. The rush to generalize information theory to all sorts of scientific areas, some of them conjectures of dubious scientific merit, led Shannon to decry this "bandwagon effect," remarking that information theory "has perhaps ballooned to an importance beyond its actual accomplishments. . . . Seldom do more than a few of nature's secrets give way at one time." [3]

Despite Shannon's disclaimer, information- and communication-based models have proved to be enormously useful in the sciences because so many important phenomena can be seen in terms of messages. Human bodies can be better understood as complex communication networks than as clockwork-like machines. The error-correcting codes guaranteed by Shannon's "noisy channel" theorem are just as useful for genetic control of protein synthesis as for communication protocols in a computer network. Shannon's MIT colleague, Noam Chomsky, has used similar tools in his exploration of the "deep structure" of language.[4]

With all these higher-level abstractions, Shannon did not abandon all thought of the potential of digital computers. Where Wiener saw the computer as a self-controlling mechanism and von Neumann saw a device with logical as well as mathematical properties, Shannon tended to think of ENIAC and UNIVAC as information processing machines.

Like Turing and other mathematicians since then, Shannon was fascinated with the idea that something as sophisticated and essentially human as chess playing could, in theory, be emulated by some future version of these devices. In February, 1950, Shannon published "A Chess Playing Machine" in *The Scientific American*. Half a decade before anyone dared to name the endeavor "artificial intelligence research," Shannon pointed out what a very few people then recognized—that electronic digital computers could "be adapted to work symbolically with elements representing words, propositions or other conceptual entities."

A chess game is a Turing machine. And a universal Turing machine, given the properly coded rules, ought to be able to play chess. Shannon pointed out that the way most people would design a machine to play chess—to mechanically examine each alternative move and evaluate it, the so-called brute-force method—would be virtually impossible, even on the fastest imaginable computer. He estimated that a typical chess game has about 10^{120} possible moves, so "A machine calculating one variation each millionth of a second would require over 10^{95} years to decide on its first move!"

This "combinatorial explosion"—the rapid and overwhelming buildup of alternatives in any system in which each level leads to two or more deeper levels, which lead to two or more deeper levels—was another one of those secrets of nature that Claude Shannon was in the habit of turning up. The explosive expansion of the number of alternative decisions is a barrier that confronts any attempt to exhaustively examine a branching structure, and continues to confront programmers who seek to emulate cognitive functions by performing searches through problem spaces.

Turing and Shannon were altogether serious in their interest in chess, because of the complexity of the game in relation to the simplicity of its rules, and because they suspected that the shortcut needed to perform this kind of time-consuming search-procedure would also be a clue to the way brains solved all sorts of problems.

A chess playing program was also interesting because it was a relative of the kind of informational entities known as *automata* that von Neumann and Turing had been toying with. Once again, like Turing's universal machines, these automata were theoretical devices that did not exist at that time, but were possible to build, in principle. For years, Shannon experimented with almost absurdly simple homemade versions—mechanical mice that were able to navigate simple mazes.

In 1953, Shannon wrote a paper, "Computers and Automata," in which he posed questions that continue to be of acute interest to psychologists as well as computerists.[5] Can a chess playing program learn from

Claude Shannon in his Bell Laboratories days, putting one of his mechanical mice into a maze for testing. (Courtesy of AT&T Bell Laboratories.)

its mistakes? Is it possible to build a machine that can diagnose itself and repair its own malfunctions? Can computer programs ("virtual machines") be created that enable computers to write their own software to the specifications of the human user? Can the way human brains process information (known in some hard-core AI circles as "wetware") ever be effectively simulated by hardware and software?

In the summer of 1953, while he was working on these ideas, Shannon hired two temporary laboratory assistants named Minsky and McCarthy, another pair of prodigies who knew some fancy mathematics and thought they could do big things with computers. Here were the first members of the first native generation of computer scientists, the ones who already knew about electronics and cybernetics and information theory and brain physiology and were looking for something ambitious to do with it all. They ended up in the right place when they dug up Shannon in the midst of Bell Laboratories.

Shannon had long spoken of his suspicion that the future evolution of more sophisticated computer hardware would make it possible to construct software capable of simulating some parts of human cognition. But these younger guys were blatant believers. They were out to build an intelligence, and didn't mind saying so. McCarthy and Shannon edited a book on automata, and three years later, in 1956, Shannon joined Minsky, McCarthy, and an IBM computer researcher, Nathaniel Rochester, in sponsoring a summer conference at Dartmouth University, to set goals for this new field. The new field they gathered to discuss was a branch of science that did not yet have a name, but which was founded on the assumption that the existence of computers now made it possible to consider creating an artificial version of the most complex system known to science—human intelligence.

It was around 1956 that McCarthy started using the words "artificial intelligence." The Dartmouth Conference was the constitutional convention of the artificial intelligence faction, and it was also the place where two virtually unknown Rand programmers named Alan Newell and Herbert Simon breezed in from Santa Monica with a piece of software they wrote with Cliff Shaw. To everyone's astonishment, it was a program—the famous *Logic Theorist* that could prove theorems from Russell and Whitehead's *Principia Mathematica*—that actually did what the rest of them thought they were there to plan to do.

Hopes were high for the AI rebels in 1956 and 1957. Major efforts were under way and ambitious goals were in sight. A very few unorthodox thinkers staked their careers on the conviction that this branch of computer science, formerly a branch of science fiction, would soon be seen as more important than anything else humankind had ever attempted: Minsky remained at MIT and concentrated on the problem of how knowledge is represented in minds and machines; Newell and Simon (now a Nobel laureate) began their long association with one another and with Carnegie-Mellon University, where they concentrated on the information processing approach to psychology and AI design; McCarthy created LISP, a language specifically for conducting AI research, and left MIT to preside over Stanford's AI laboratory.

Claude Shannon went back to his chess playing machines and continued building the mechanical mice that could learn how to run simple mazes. In 1956, Robert Fano, the electrical engineering student who witnessed Norbert Wiener's "Entropy is information!" exclamations back in the summer of 1947, brought Shannon to MIT from Bell Laboratories.

His professional standing was so far beyond reproach that his occasional unicycle excursions through MIT halls, and his reluctance to lecture or publish frequently, hardly dented Shannon's reputation. In fact, his reputation had reached such mythological proportions that he had to start writing disclaimers. Fame wasn't something he wanted or needed. By 1960, he didn't even come into his office.

In the 1960s, Shannon became interested in the stock market as a real-world experiment in probability theory, and rumor has it that he didn't do too badly. He began to seriously extend his analysis of communications and messages to the English language. Nobody but Shannon knows the full extent of his discoveries. Robert Fano (who went on to become the administrative director of Project MAC) recently said this of Shannon: [6]

> There is a significant body of work he did in the 1950s that has never been printed. He doesn't want someone else to write his papers for him, and he won't write them himself. It's as simple and as complicated as that. He doesn't like to teach. He doesn't like giving lectures. His lectures are beautiful, jewels all of them. They sound spontaneous, but in reality they are very, very carefully prepared.

In the early sixties, one of the extremely few students Shannon personally took on, another MIT-bred prodigy by the name of Ivan Sutherland, made quite a splash on the computer science scene. By the mid-1970s, Shannon, now in his sixties, had become a literal gray eminence. By the early 1980s, he still hadn't stopped thinking about things, and considering his track record, it isn't too farfetched to speculate that his most significant discoveries have yet to be published.

In the late 1950s, around the same time that Shannon began to retreat from public life, the artificial intelligence pioneers began to stake out ambitious territories for their laboratories—goals like automatic theorem-proving programs, or knowledge representation languages, or robotics—and it began to be possible to dream of computers that could be used as laboratories for running experiments in new kinds of AI programs. Then fate put a little pressure on the story once again.

This time, it was not a war, but an implicit threat of war. The space race and the computer revolution were ready to be launched by 1957, and the information processing devices pioneered by the World War II creators of computing were ready to leave the laboratories and begin to infiltrate the real world. As usual, things started popping when an MIT professor stumbled onto something big.

CHAPTER SEVEN

Machines to Think With

In the spring of 1957, while he continued to carry out the duties of an MIT researcher and professor, Dr. J.C.R. Licklider noted every task he did during the day and kept track of the time he spent on each one. He didn't know it then, but that unofficial experiment prepared the way for the invention of interactive computing—the technology that bridged yesteryear's number crunchers and tomorrow's mind amplifiers.

Licklider's research specialty was *psychoacoustics*. During World War II, he had explored ways electronics could be applied to understanding human communications. Specifically, he wanted to learn how the human ear and brain are able to convert atmospheric vibrations into the perception of distinct sounds. After the war, MIT was the center of a number of different attempts to use electronic mechanisms to model parts of the nervous system—a movement in biology and psychology as well as engineering that was inspired by the work of Wiener and others in the interdisciplinary field of cybernetics. Licklider was one of the researchers attracted to this paradigm, not strictly out of the desire to build a new kind of machine, but out of the need for new ways to simulate the activities of the human brain. This need, inspired by cybernetics, extended simultaneously into engineering and physiology. Computers were the last thing on Licklider's mind—until his theoretical models of human perceptual mechanisms got out of hand.

By the late 1950s, Licklider was trying to build mathematical and electronic models of the mechanisms the brain uses to process the perception of sounds. Part of the excitement generated during the early days of cybernetic research came from the prospect of studying mechanical models of living organisms to help create theoretical models of the way those organisms function, and vice versa. Licklider thought he might be onto a good idea with an intricate neural model of pitch perception, but quickly learned, to his dismay, that his mathematical model had grown too complex to work out by hand in a reasonable length of time, even by using the analog computers that were then available. And until the mathematical model could be worked out, there was no hope for progress in building a mechanical model of pitch perception.

The idea of building a mathematical or electronic model was meant to *simplify* the task of understanding the complexities of the brain, like plotting a graph to see the key relationships in a collection of data. But the models themselves now began to grow unmanageably complex. Like Mauchly with his meteorological data, twenty years before, Licklider found that he was spending more and more of his time dealing with the calculations he needed to do to create his models, which left less time for what he considered to be his primary occupation—thinking about what all that information meant. Beneath those numbers and graphs was his real objective—the theoretical underpinnings of human communication.

Although he was primarily interested in how the brain processes auditory information, he felt that he was spending most of his time putting things into files or taking them out, as well as managing the increasing amounts of numerical data he needed to construct the models he had in mind. Out of curiosity, he wondered if any of his colleagues had looked into the way scientific researchers spent their time.

When he couldn't find any time-and-motion studies on information-shuffling researchers like himself, Licklider decided to keep track of his own activities as he went through his normal working day. "Although I was aware of the inadequacy of the sampling," he later wrote, with the modesty and humor that he is known for among his colleagues, "I served as my own subject." [1]

It didn't take long to discover that his main occupation, even when he wasn't keeping records of his behavior, was centered on keeping records of everything else. Astonishing as it must have seemed to any self-respecting scientist like himself, his observations revealed that about 85 percent of his "thinking" time was actually spent "getting into a position to think, to make a decision, to learn something I needed to know. Much more time went into finding or obtaining information than into digesting it." [2]

Like almost any other experimentalist, he couldn't begin to make sense of the psychoacoustic data until he could see it translated into the form of graphs. Plotting the graphs took days. Even teaching his assistants how to plot graphs took hours. As soon as the graphs were finished and he was able to look at them, the relationships he was seeking became immediately obvious. It was grossly inefficient and tedious to spend days plotting graphs that took seconds to interpret.

While he had always thought of interpretation and evaluation as his most important functions as a scientist, Licklider's analysis of his research behavior showed that most of his tasks were clerical or mechanical: "searching, calculating, plotting, transforming, determining the logical or dynamic consequences of a set of assumptions or hypotheses, preparing the way for a decision or an insight. Moreover, my choices of what to attempt and what not to attempt were determined to an embarrassingly great extent by considerations of clerical feasibility, not intellectual capability." [3]

The conclusion he reached, while it doesn't sound so radical today, was shocking when it occurred to him in 1957. A less modest man might not have been able to bring himself to face the conclusion: Licklider decided, on the basis of his informal self-study, that most of the tasks that take up the time of any technical thinker would be performed more effectively by machines.

This was a thought that was occurring to one or two other people at about the same time—notably Doug Engelbart, out in California. But because of his association with certain military-sponsored research projects conducted at MIT in the 1950s, there was an important difference between Licklider and the others who dreamed of converting computers into some kind of mind-amplifying tool. This crucial difference was the fact that Licklider had reached his conclusion not long before circumstances put him at the center of power in the one institution capable of sponsoring the creation of an entire new technology.

At that point in the history of computer technology—a field in which Licklider had been only tangentially involved until then—no respectable computer scientist would dare suggest that computer technology ought to be totally revamped so that scientists could use these machines to help keep track of data and build theoretical models of the phenomena they were studying. To those who were wild enough to make such a suggestion—especially the young MIT computer mavericks who were founding the field of artificial intelligence around that time—the idea might have seemed too obvious and too trivial to pursue. In any case, the AI founders were more interested in replacing the scientist than the scientist's file clerk. Licklider, however, was neither a respectable computer scientist nor a computer maverick, but a psychologist with some expertise in electronics. And like any other competent investigator, he followed where the data led him.

In the late 1950s, Licklider had no real expertise in digital computer design, and although he knew that only a computer could give him what he needed, he didn't think that the kinds of computers then available, and the kinds of things they did, were suitable for building a sort of "electronic file clerk." He knew that *data processing* wasn't what he wanted.

If you were the Census Office, overflowing with information on a couple of hundred million people, and for some crazy reason you wanted to find out how many divorced people over sixty lived on farms in the sun belt, you could use a UNIVAC to perform the sorting and calculating needed to tell you what you wanted to know. That was data processing. If you had a payroll for 10,000 employees to calculate every other Friday and needed to transform time sheets into entries in a ledger and print up all the checks—data processing power was just what you could buy from your local IBM representative.

Data processing involved certain constraints on *what* could be done with computers, and constraints on *how* one went about doing these things. Payrolls, mathematical calculations, and census data were the proper kinds of tasks. An arcane process known as "batch processing" was the proper way to do these things. If you had a problem to solve, you had to encode your program and the data the program was meant to operate upon, usually in one of the two major computer languages—FORTRAN and COBOL. The encoded program and data were converted into boxes full of what had become universally known as "IBM cards"—the kind you aren't supposed to spindle, fold, or mutilate. The cards were delivered to a systems administrator at the campus "computer center" or the corporate "data processing center." This specialist was the only one allowed to submit the program to the machine, and the person from whom you would retrieve your printout hours or days later.

But if you wanted to plot ten thousand points on a line, or turn a list of numbers into a graphic model of airflow patterns over an airplane wing, you wouldn't want data processing or batch processing. You would want *modeling*—an exotic new use for computers that the aircraft designers were then pioneering. All Licklider sought, at first, was a mechanical servant to take care of the clerical and calculating work that accompanied model building. Not long after, however, he began to wonder if computers could help *formulate* models as well as calculate them.

When he attained tenure, later that same year, Licklider decided to join a consulting firm near Cambridge named Bolt, Beranek & Newman. They offered him an opportunity to pursue his psychoacoustic research—and a chance to learn about digital computers.

"BB&N had the first machine that Digital Equipment Company made, the PDP-1," Licklider recalled in 1983. The quarter-million-dollar machine was the first of a continuing line of what came to be called, in the style of the midsixties, "minicomputers." Instead of costing millions of dollars and occupying most of a room, these new, smaller, powerful computers only cost hundreds of thousands of dollars, and took up about the same amount of space as a couple of refrigerators. But they still required experts to operate them. Licklider therefore hired a research assistant, a college dropout who was knowledgeable about computers, an exceptionally capable young fellow by the name of Ed Fredkin, who was later to become a force in artificial intelligence research—the first of many exceptionally capable young fellows who would be drawn to Licklider's crusade to build a new kind of computer and create a new style of computing.

Fredkin and others at BB&N had the PDP-1 set up so that Licklider could directly interact with it. Instead of programming via boxes of punch cards over a period of days, it became possible to feed the programs and data to the machine via a high-speed paper tape; it was also possible to change the paper tape input *while the program was running*. The operator could *interact* with the machine for the first time. (The possibility of this kind of interaction was duly noted by a few other people who turned out to be influential figures in computer history. A couple of other young computerists at MIT, John McCarthy and Marvin Minsky, were also using a PDP-1 in ways computers weren't usually used.)

The PDP-1 was primitive in comparison with today's computers, but it was a breakthrough in 1960. Here was the model builder that Licklider had first envisioned. This fast, inexpensive, interactive computer was beginning to resemble the kind of device he dreamed about back in his psychoacoustic lab at MIT, when he first realized how his ability to theorize always seemed constrained by the effort it took to draw graphs from data.

"I guess you could say I had a kind of religious conversion," Licklider admits, remembering how it felt, a quarter of a century ago, to get his hands on his first interactive computer. As he had suspected, it was

Professor J.C.R. Licklider with some precomputer electronic equipment, several years before his ARPA days. (Courtesy of the MIT Museum.)

indeed possible to use computers to help build models from experimental data and to help make sense of any complicated collection of information.

Then he learned that although the computer was the right *kind* of machine he needed to build his models, even the PDP-1 was hopelessly crude for the phenomena he wanted to study. Nature was far too complicated for 1960-style computers. He needed more memory components and faster processing of large amounts of calculations. As he began to think about the respective strengths and deficiencies of computers and brains, it occurred to him that what he was seeking was an alternative to the human-computer relationship as it then existed.

Since the summer of 1956, when they met at Dartmouth to define the field, several young computer and communications scientists Licklider knew from MIT had been talking about a vaguely distant future when machines would surpass the limits of human intelligence. Licklider was more concerned with the shorter-term potential of computer-human relations. Even at the beginning, he realized that technical thinkers of every kind were starting to run up against the problems he had started noticing in 1957. Let the AI fellows worry about ways to build chess playing or language translating machines. What he and a lot of other people needed was an intelligent assistant.

Although he was convinced by his "religious conversion to interactive computing"—a phrase that has been used over and over again by those who participated in the events that followed—Licklider still knew too little about the economics of computer technology to see how it might become possible to actually construct an intelligent laboratory assistant. Although he didn't know how or when computers would become powerful enough and cheap enough to serve as "thinking tools," he began to realize that the general-purpose computer, if it was set up in such a way that humans could interact with it directly, could evolve into something entirely different from the data processors and number crunchers of the 1950s. Although the possibility of a creating a personal tool still seemed economically infeasible, the idea of modernizing a community-based resource, like a library, began to appeal to him. He got fired up about the idea Vannevar Bush had mentioned in 1945, the concept of a new kind of library to fit the world's new knowledge system.

"The PDP-1 opened me up to ideas about how people and machines like this might operate together in the future," Licklider recalled in 1983, "but I never dreamed at first that it would ever become economically feasible to give everybody their own computer." It did occur to him that these new computers were excellent candidates for the super-mechanized libraries that Vannevar Bush had prophesied. In 1959, he wrote a book entitled *Libraries of the Future*, describing how a computer-based system might create a new kind of "thinking center."

The computerized library as he first described it in his book did not involve anything as extravagant as giving an entire computer to every person who used it. Instead he described a setup, the technical details of which he left to the future, by which different humans could use remote extensions of a central computer, all at the same time.

After he wrote the book, during the exhilarating acceleration of research that began in the post-Sputnik era, Licklider discovered what he and others who were close to developments in electronics came to call "the rule of two": Continuing miniaturization of its most important components means that the cost effectiveness of computer hardware doubles every two years. It was true in 1950 and held true in 1960, and beyond even the wildest imaginings of the transistor revolutionaries, it was still true in 1980. A small library of books and articles have been written about the ways this phenomenon has fueled the electronics revolution of the past three decades. It looks like it will continue to operate until at least 1990, when personally affordable computers will be *millions* of times more powerful than ENIAC.

Licklider then started to wonder about the possibility of devising something far more revolutionary than even a computerized library. When it began to dawn on him that this relentlessly exponential rate of growth would make available computers over a hundred times as powerful as the PDP-1 at one tenth the cost within fifteen years, Licklider began to think about a system that included both the electronic powers of the computer and the cortical powers of the human operator. The crude interaction between the operator and the PDP-1 might be just the beginning of a powerful new kind of human-computer partnership.

A new kind of computer would have to evolve before this higher level of human-machine interaction could be possible. The way the machine was operated by people would have to change, and the machine itself would have to become much faster and more powerful. Although he was still a novice in digital computer design, Licklider was familiar with vacuum tube circuitry and enough of an expert in the hybrid discipline of "human factors engineering" to recognize that the mechanical assistant he wanted would need capabilities that would be possible only with the ultrafast computers he foresaw in the near future.

When he began applying the methods he had been using in human factors research to the informational and communication activities of technical thinkers like himself, Licklider found himself drawn to the idea of a kind of computation that was more dynamic, more of a dialogue, more of an aid in *formulating* as well as *plotting* models. Licklider set forth in 1960 the specifications for a new species of computer and a new mode of thinking to be used when operating them, a specification that is still not fully realized, a quarter of a century later: [4]

> The information-processing equipment, for its part, will convert hypotheses into testable models and then test the models against data (which the human operator may designate roughly and identify as relevant when the computer presents them for his approval). The equipment will answer questions. It will simulate the mechanisms and models, carry out the procedures, and display the results to the operator. It will transform data, plot graphs ("cutting the cake" in whatever way the human operator specifies, or in several alternative ways if the human operator is not sure what he wants). The equipment will interpolate, extrapolate, and transform. It will convert static equations or logical statements into dynamic models so the human operator can examine their behavior. In general, it will carry out the routinizable, clerical operations that fill the intervals between decisions.
>
> In addition, the computer will serve as a statistical-inference, decision-theory, or game-theory machine to make elementary evaluations of suggested courses of action whenever there is enough basis to support a formal statistical analysis. Finally, it will do as much diagnosis, pattern matching, and relevance recognizing as it profitably can, but it will accept a clearly secondary status in those areas.

The first research in the 1950s into the use of computing equipment for assisting human control of complex systems was a direct result of the need for a new kind of air defense command-and-control system. Licklider, as a human factors expert, had been involved in planning these early air defense communication systems. Like the few others who saw this point as early as he did, he realized that the management of complexity was the main problem to be solved during the rest of the twentieth century and beyond. Machines would have to help us keep track of the complications of keeping global civilization alive and growing. And humans were going to need new ways of attacking the big problems that would result from our continued existence and growth.

Assuming that survival and a tolerable quality of existence are the most fundamental needs for all sane, intelligent organisms, whether they are of the biological or technological variety, Licklider wondered if the best arrangement for both the human and human-created symbol-processing entities on this planet might not turn out to be neither a master-slave relationship nor an uneasy truce between competitors, but a *partnership*.

Then he found the perfect metaphor in nature for the future capabilities he had foreseen during his 1957–1958 "religious conversion" to interactive computing and during those 1958–1960 minicomputer encounters that set his mind wandering through the informational ecology of the future. The newfound metaphor showed him how to apply his computer experience to his modest discovery about how technical thinkers spend their time. The idea that resulted grew into a theory so bold and immense that it would alter not only human history but human evolution, if it proved to be true.

In 1960, in the same paper in which he talked about machines that would help formulate as well as help construct theoretical models, Licklider also set forth the concept of the kind of human-computer relationship that he was later to be instrumental in initiating: [5]

> The fig tree is pollinated only by the insect *Blastophaga grossorum*. The larva of the insect lives in the ovary of the fig tree, and there it gets its food. The tree and the insect are thus heavily interdependent: the tree cannot reproduce without the insect; the insect cannot eat without the tree; together, they constitute not only a viable but a productive and thriving partnership. This cooperative "living together in intimate association, or even close union, of two dissimilar organisms" is called symbiosis.
>
> "Man-computer symbiosis" is a subclass of man-machine systems. There are many man-machine systems. At present, however, there are no man-computer symbioses. . . . The hope is that, in not too many years, human brains and computing machines will be coupled together very tightly, and that the resulting partnership will think as no human being has ever thought and process data in a way not approached by the information-handling machines we know today.

The problems to be overcome in achieving such a partnership were only partially a matter of building better computers and only partially a matter of learning how minds interact with information. The most important questions might not be about either the brain or the technology, but about the way they are coupled.

Licklider, foreseeing the use of computers as tools to build better computers, concluded that 1960 would begin a transitional phase in which we humans would begin to build machines capable of learning to communicate with us, machines that could eventually help us to communicate more effectively, and perhaps more profoundly, with one another.

By this time, he had strayed far enough off the course of his psychoacoustic research to be seduced by the prospect of building the device he first envisioned as a tool to help him make sense of his laboratory data. Like Babbage, who needed a way to produce accurate logarithm tables, or Goldstine, who wanted better firing tables, or Turing, who wanted a perfectly definite way to solve mathematical and cryptological problems, Licklider began to move away from his former goals as he got caught up in the excitement of creating the tools he needed.

Except Licklider wasn't an astronomer and tinkerer like Babbage, a ballistician like Goldstine, or a mathematician and code-breaker like Turing, but an experimental psychologist with some practical electronic experience. He had set out to build a small model of one part of human awareness—pitch perception—and ended up dreaming about machines that could help him think about models.

As other software visionaries before and after him knew very well, Licklider's vision, as grandiose as it might have been, wasn't enough in itself to ensure that anything would ever happen in the real world. An experimental psychologist, even an MIT professor, is hardly in a position to set armies of computer engineers marching toward an interactive future. Like von Neumann and Goldstine meeting on the railroad platform at Aberdeen, or Mauchly and Eckert encountering each other in an electronics class at the Moore School, Licklider happened upon his destiny through accidental circumstances, because of time he spent at a place called "Lincoln Laboratory," an MIT facility for top-secret defense research, where he was a consultant during a critical transition period in the history of information processing.

It was his expertise in the psychology of human-machine interaction that led Licklider to a position where he could make big things out of his dreams. In the early and mid 1950s, MIT and IBM were involved in building what were to become the largest computers ever built, the IBM AN/FSQ-7, as the control centers of a whole new continental air defense system for the United States. SAGE (Semi-Automatic Ground Environment) was the Air Force's answer to the new problem of potential nuclear bomber attack. The computers weighed three hundred tons, took up twenty thousand feet of floor space, and were delivered in eighteen large vans apiece. Ultimately, the Air Force bought fifty-six of them.

MIT set up Lincoln Laboratory in Lexington, Massachusetts, to design SAGE. At the other end of the continent, System Development Corporation in Santa Monica (the center of the aircraft industry) was founded to create the software for SAGE. Some of the thorniest problems that were encountered on this project had to do with devising ways to make large amounts of information available in human-readable form, quickly enough for humans to make fast decisions about that information. It just wouldn't do for your computers to take three days to evaluate all the radar and radio-transmitted data before the Air Defense Command could decide whether or not an air attack was under way.

Some of the answers to these problems were formulated in the "Whirlwind" project at the MIT computing center, where high-speed calculations were combined with computer controls that resembled aircraft controls. Other answers came from specialists in human perception (like Licklider), who devised new ways for computers to present information to people. With the exception of the small crew of the earlier Whirlwind project, SAGE operators were the first computer users who were able to see information on visual display screens; moreover, operators were able to use devices called "lightpens" to alter the graphic displays by touching the screens. There was even a primitive decision-making capacity built into the system: the computer could suggest alternate courses of action, based on its model of the developing situation.

The matter of display screens began to stray away from electronics and into the area of human perception and cognition, which was Licklider's cue to join the computer builders. But even before Lincoln Laboratory was established in 1953–1954, Licklider had been consulted about the possibility of developing a new technology for displaying computer information to human operators for the purposes of improving air defense capabilities. Undoubtedly, the seeds of his future ideas about human-computer symbiosis were first planted when he and other members of what was then called "the presentation group" considered the kinds of visual displays air defense command centers would need.

The presentation group was where he first became acquainted with Wesley Clark, one of MIT's foremost computer builders. Clark had been a principal designer of Whirlwind, the most advanced computer system to precede the SAGE project. Whirlwind, the purpose of which was to act as a kind of flight simulator, was in many ways the first hardware ancestor of the personal computer, because it was designed to be operated by a single "test pilot." It was also used for modeling aerodynamic equations. While it was only barely interactive in the sense that Licklider desired, Whirlwind was the first computer fast enough to solve aerodynamic equations in "real time"—as the event that was being modeled was actually happening. Real-time computation was not only a practical necessity for the increasingly complicated job of designing high-speed jet aircraft; it was a necessary prerequisite for creating the guidance systems of rockets, the technological successors to jet aircraft.

Ironically, by the time SAGE became fully operational in 1958, the entire concept of ground-based air defense against bomber attack had been made obsolete on one shocking day in October, 1957, when a little beeping basketball by the odd name of "Sputnik" jolted the American military, scientific, and educational establishments into a frenzy of action. The fact that the Russians could put bombs in orbit set off the most intensive peacetime military research program in history. When the Soviets repeated their triumph by putting Yuri Gagarin into space, a parallel impetus started the U.S. manned space effort on a similar course.

In the same way that the need for ballistics calculations indirectly triggered the invention of the general-purpose digital computer, the aftermath of Sputnik started the development of interactive computers, and eventually led directly to the devices now known as personal computers. Just as von Neumann found himself in the center of political-technological events in the ENIAC era, Licklider was drawn into a central role in what became known at "the ARPA era."

The "space race" caused a radical shakeup in America's defense research bureaucracy. It was decided at the highest levels that one of the factors holding up the pace of space-related research was the old, slow way of evaluating research proposals by submitting them for anonymous review by knowledgeable scientists in the field (a ritual known as "peer review" that is still the orthodox model for research funding agencies).

The new generation of Camelot-era whiz kids from the think tanks, universities, and industry, assembled by Secretary McNamara in the rosier days before Vietnam, were determined to use the momentum of the post-Sputnik scare to bring the Defense Department's science and technology bureaucracy into the space age. Something had to be done to streamline the process of technological progress in fields vital to the national security. One answer was NASA, which grew from a tiny subagency to a bureaucratic, scientific, and engineering force of its own. And the Defense Department created the Advanced Research Projects Agency, ARPA. ARPA's mandate was to find and fund bold projects that had a chance of advancing America's defense-related technologies by orders of magnitude—bypassing the peer review process by putting research administrators in direct contact with researchers.

Because of their involvement with previous air defense projects, a few of Licklider's friends from Lincoln, like Wesley Clark, were involved in the changeover to the fast-moving, forward-thinking, well-funded, results-oriented ARPA way of doing things. Clark designed the TX-0 and TX-2 computers at MIT and Lincoln. The first of these machines became famous as the favorite tool of the "hackers" in "Building 26," who later became the legendary core of Project MAC. The second machine was designed expressly for advanced graphic display research.

Graphic displays were esoteric devices in 1960, known only to certain laboratories and defense facilities. Aside from the PDP-1, almost every computer displayed information via a teletype machine. But there was an idea floating around Lincoln that SAGE-like displays might be adapted to many kinds of computers, not just the big ones used to monitor air defenses. By 1961, the psychology of graphic displays had become something of a specialty for Licklider. Between BB&N and Lincoln, he was spending more time with electrical engineers than with psychologists.

Through his computer-oriented colleagues, Licklider became acquainted with Jack Ruina, director of ARPA in the early 1960s. Ruina wanted to do something about computerizing military command and control systems on all levels—not just air defense—and wanted to set up a special office within ARPA to develop new information processing techniques. ARPA's goal was to leapfrog over conventional research and development by funding attempts to make fundamental breakthroughs. And Licklider's notion of creating a new kind of computer capable of directly interacting with human operators via a keyboard and display screen interface (instead of relying on batch processing or even paper-tape input) convinced Ruina that the minority of computer researchers Licklider was talking about might just lead to such a possible breakthrough.

"I got Jack to see the pertinence of interactive computing, not only to military command and control, but to the whole world of day-to-day business," Licklider recalls. "So, in October, 1962 I moved into the Pentagon and became the director of the Information Processing Techniques Office." And that event, as much as any other development of that era, marked the beginning of the age of personal computing.

The unprecedented technological revolution that began with the post-Sputnik mobilization and reached a climax with Neil Armstrong's first step on the moon a little more than a decade later was in very large part made possible by a parallel revolution in the way computers were used. The most spectacular visual shows of the space age were provided by the enormous rockets. The human story was concentrated on the men in the capsules atop the rockets. But the unsung heroics that ensured the success of the space program were conducted by men using new kinds of computers.

Remember the crew at mission control, who burst into cheers at a successful launch, and who looked so cool nineteen hours later when the astronaut and the mission depended on their solutions to unexpected glitches? When the bright young men at their computer monitors were televised during the first launches from Cape Canaveral, the picture America saw of their working habitat reflected the results of the research Licklider and the presentation group had performed. After all, the kinds of computer displays you need for NORAD (North American Air Defense Command) aren't too different from the kind you need for NASA—in both cases, groups of people are using computers to track the path of multiple objects in space. NASA and ARPA shared results in the computer field—a kind of bureaucratic cooperation that was relatively rare in the pre-Sputnik era.

Because the Russians appeared to be far ahead of us in the development of huge booster rockets, it was decided that the United States should concentrate on guidance systems and ultralight (i.e., ultraminiature) components for our less powerful rockets—a policy that was rooted in the fundamental thinking established by the ICBM Committee a few years back, in the von Neumann days. Therefore, the space program and the missile program both required the rapid development of very small, extremely reliable computers.

The decision of the richest, most powerful nation in history to put a major part of its resources into the development of electronic-based technologies happened at an exceptionally propitious moment in the history of electronics. The basic scientific discoveries that made the miniaturization revolution possible—the new field of semiconductor research that produced the transistor and then the integrated circuit—made it clear that 1960 was just the beginning of the rapid evolution of computers. The size, speed, cost, and energy requirements of the basic switching elements of computers changed by orders of magnitude when electron tubes replaced relays in the late 1940s, and again when transistors replaced tubes in the 1950s, and now integrated circuits were about to replace transistors in the 1960s. In the blue-sky labs, where the engineers were almost outnumbered by the dreamers, they were even talking about "large-scale integration."

When basic science makes breakthroughs at such a pace, and when technological exploitation of those discoveries is so deliberately intensified, a big problem is in being able to envision *what's possible and preferable to do next*. The ability to see a long range goal, and to encourage the right combination of boldness and pragmatism in all the subfields that could contribute to achieving it, was the particular talent that Licklider brought onto the scene. And with Licklider came a new generation of designers and engineers and programmers who had their sights on something the pre-Sputnik computer orthodoxy would have dismissed as science fiction. Suddenly, human-computer symbiosis wasn't an esoteric hypothesis in a technical journal, but a national goal.

When Licklider went to ARPA, he wasn't given a laboratory, but an office, a budget, and a mandate to raise the state of the art of information processing. He started by supporting thirteen different research groups around the country, primarily at MIT; System Development Corporation (SDC); the University of California at Berkeley, Santa Barbara, and Los Angeles; USC; Rand; Stanford Research Institute (now SRI International); Carnegie-Mellon University; and the University of Utah. And when his office decided to support a project, that meant providing thirty or forty times the budget that the researchers were accustomed to, along with access to state-of-the-art research technology and a mandate to think big and think fast.

A broad range of new capabilities that Licklider then called "interactive computing" was the ultimate goal, and the first step was an exciting new concept that came to be known as *time-sharing*. Time-sharing was to be the first, most important step in the transition from batch processing to the threshold of personal computing (i.e, one person to one machine). The idea was to create computer systems capable of interacting with many programmers at the same time, instead of forcing them to wait in line with their cards or tapes.

Exploratory probes of the technologies that could make time-sharing possible had been funded by the Office of Naval Research and the Air Force Office of Scientific Research before ARPA stepped in. Licklider beefed up the support to the MIT Cambridge laboratory where AI researchers were working on their own approach to "multiaccess computing." Project MAC, as this branch became known, was the single node in the research network where AI and computer systems design were, for a few more years, cooperative rather than divergent.

MAC generated legends of its own, from the pioneering AI research of McCarthy, Minsky, Papert, Fredkin, and Weizenbaum, to the weird new breed of programmers who called themselves "hackers," who held late-night sessions of "Spacewar" with a PDP-1 they had rigged to fly simulated rockets around an oscilloscope screen and shoot dots of light at one another. MAC was one of the most important meeting grounds of both the AI prodigies of the 1970s and the software designers of the 1980s. By the end of the ARPA-supported heyday, however, the AI people and the computer systems people were no longer on the same track.

One of Licklider's first moves in 1962–1963 was to set up an MIT and Bolt, Beranek & Newman group in Massachusetts to help Systems Development Corporation out in Santa Monica in producing a transistorized version of the SAGE-based time-sharing prototypes, which were based on the old vacuum-tube technology. The first step was to get a machine to all the researchers that was itself interactive enough that it could be used to design more interactive versions—the "bootstrapping" process that became the deliberate policy of Licklider and his successors. The result was that university laboratories and think tanks around the country began to work on the components of a system that would depend on engineering and software breakthroughs that hadn't been achieved yet.

The time-sharing experience turned out to be a cultural as well as a technological watershed. As Licklider had predicted, these new tools changed the way information was processed, but they also changed the way people thought. A lot of the researchers who were to later participate in the creation of personal computer technology got their first experience in the high-pressure art and science of interactive computer design in the first ARPA-funded time-sharing projects.

One of the obstacles to achieving the kind of interactive computing that Licklider and his growing cadre of "converts" envisioned lay in the slowness and low capacity of the memory component of 1950-style computers; this hardware problem was partially solved when Jay Forrester, director of the Whirlwind project, came up with "magnetic core memory." The advent of transitorized computers promised even greater memory capacity and faster access times in the near future. A different problem, characterized by the batch-processing bottleneck, stemmed from the way computers were set up to accept input from human operators; a combination of hardware and software innovations were converging on a direct keyboard-to-computer input.

Another one of the obstacles to achieving the overall goal of interactive computing lay not in the way the computer processed information—an issue that was addressed by the time-sharing effort—but in the primitive way computers were set up to display information to human operators. Lincoln Laboratory was the natural place to concentrate the graphics effort. Another graphics-focused effort was started at the University of Utah. The presentation group veterans, expanded by the addition of experts in the infant technology of transistor-based computer design, began to work intensively on the problem of display devices.

Licklider remembers the first official meeting on interactive graphics, where the first wave of preliminary research was presented and discussed in order to plan the assault on the main problem of getting information from the innards of the new computers to the surface of various kinds of display screens. It was at this meeting, Licklider recalls, that Ivan Sutherland first took the stage in a spectacular way.

"Sutherland was a graduate student at the time," Licklider remembers, "and he hadn't been invited to give a paper." But because of the graphics program he was creating for his Ph.D. thesis, because he was a protégé of Claude Shannon, and because of the rumors that he was just the kind of prodigy ARPA was seeking, he was invited to the meeting. "Toward the end of one of the last sessions," according to Licklider, "Sutherland stood up and asked a question of one of the speakers." It was the kind of question that indicated that this unknown young fellow might have something interesting to say to this high-powered assemblage.

So Licklider arranged for him to speak to the group the next day: "Of course, he brought some slides, and when we saw them everyone in the room recognized his work to be quite a lot better than what had been described in the formal session." Sutherland's thesis, a program developed on the TX-2 at Lincoln, demonstrated an innovative way to handle computer graphics—and a new way of commanding the operations of computers. He called it *Sketchpad*, and it was clearly evident to the assembled experts that he had leaped over their years of research to create something that even the most ambitious of them had not dared.

Sketchpad allowed a computer operator to use the computer to create, very rapidly, sophisticated visual models on a display screen that resembled a television set. The visual patterns could be stored in the computer's memory like any other data, and could be manipulated by the computer's processor. In a way, this was a dramatic answer to Licklider's quest for a fast model-builder. But Sketchpad was much more than a tool for creating visual displays. It was a kind of simulation language that enabled computers to translate abstractions into perceptually concrete forms. And it was a model for totally new ways of operating computers; by changing something on the display screen, it was possible, via Sketchpad, to change something in the computer's memory.

"If I had known how hard it was to do, I probably wouldn't have done it," Alan Kay remembers Sutherland saying about his now-legendary program. Not only was the technical theory bold, innovative, and sound, but the program actually worked. With a lightpen, a keyboard, a display screen, and the Sketchpad program running on the relatively crude real-time computers available in 1962, anyone could see for themselves that computers could be used for something else beside data processing. And in the case of Sketchpad, seeing was truly believing.

When he left ARPA in 1964, Licklider recommended Sutherland as the next director of the IPTO. "I had some hesitance about recommending someone so young," remembers Licklider, "but Bob Sproull, Ruina's successor as ARPA director, said he had no problem with his youth if Sutherland was really as bright as he was said to be." By that time, Sutherland, still in his early twenties, had established a track record for himself doing what ARPA liked best—racing ahead of the technology to accomplish what the orthodoxy considered impossible or failed to consider altogether.

When Sutherland took over, the various time-sharing, graphics, AI, operating systems, and programming language projects were getting into full swing, and the office was growing almost as fast as the industries that were spinning off the space-age research bonanza. Sutherland hired Bob Taylor, a young man from the research funding arm of NASA, to be his assistant, and ultimately his successor when he left IPTO in 1965. Licklider went to the IBM research center in 1964, then back to MIT to take charge of Project MAC in 1968.

In 1983, over a quarter century since the spring day he decided to observe his own daily activities, Licklider is still actively counseling those who build information processing technologies. After three decades of direct experience with "the rule of two," he is not sure that information engineers have even approached the physical limits of information storage and processing.

One thing scientists and engineers know now that they didn't know when he and the others started, Licklider points out, is that "Nature is very much more hospitable to information processing than anybody had any idea about in the 1950s. We didn't realize that molecular biologists had provided an existence proof for a fantastically efficient, reliable, information processing mechanism—the molecular coding of the human genetic system. The informational equivalent of the world's entire fund of knowledge can be stored in less than a cubic centimeter of DNA, which tells us that we haven't begun to approach the physical limits of information processing technology."

The time-sharing communities, and the network communities that followed them, were part of another dream—the prospect of computer-mediated communities throughout the world, extending beyond the computer experts to thinkers, artists, and business people. Licklider believes it is entirely possible that the on-line, interactive human-computer community he dreamed about will become technologically feasible sometime within the next decade. He knew all along that the frameworks of ideas and the first levels of hardware technology achieved in the 1960s and 1970s were only the foundation for a lot of work that remained to be done.

When the bootstrapping process of building better, cheaper, experimental interactive information processing systems intersects with the rising curve of electronic capabilities, and the dropping curve of computation costs, it will become possible for millions, rather than a thousand or two, to experience the kind of information environment the ARPA-sponsored infonauts knew.

In the early 1980s, millions of people already own personal computers that will become obsolete when versions a hundred times as fast with a thousand times the memory capacity come along at half of today's prices. When tens of millions of people get their hands on powerful enough devices, and a means for connecting them, Licklider still thinks the job will only be in its beginning stages.

Looking toward the day when the "intergalactic network" he speculated about in the midsixties becomes feasible, he remains convinced that the predicted boost in human cultural capabilities will take place, but only after enough people use an early version of the system to think up a more capable system that *everybody* can use: "With a large enough population involved in improving the system, it will be easier for new ideas to be born and propagated," he notes, perhaps remembering the years when interactive computing was considered a daring venture by a bunch of mavericks. The most significant issue, he still believes, is whether the medium will become truly *universal*.

"What proportion of the total population will be able to join that community? That's still an important question," Licklider concludes, still not sure whether this new medium will remain the exclusive property of a smaller group who might end up wielding disproportionate power over others by virtue of their access to these tools, or whether it will become the property of the entire culture, like literacy.

CHAPTER EIGHT

Witness to Software History: The Mascot of Project MAC

When he tried the doorknob and found it unlocked, then opened the door to Building 26 and poked his head into a room full of weirdos having a high old time with candy bars and computer programs, David Rodman knew he had discovered something. The year was 1960. David Rodman was ten years old. And 1960 was still at least four years too early for weird people to be anything but a rarity, even on college campuses.

It turned out that these pasty-faced, hollow-eyed, jargon-spewing, insanely cackling young men were the first, founding generation of dropout programming wizards to call themselves "hackers," and Building 26 was where the hotshot hired programmers of MIT's artificial intelligence Project MAC were caged until they all moved to the ninth floor of 545 Technology Square, in the early sixties.

Technology Square was MIT's space-age temple of sci-tech. The geographical move from outpost to the pinnacle of the technohierarchy reflected an elevation in the importance of the whole field of man-machine systems. MAC was set up originally by Licklider, later administered at various times by Fano, Minsky, and Papert, and the ambiguity about the meaning of the acronym was deliberate. On the level of the hackers' employers, it meant both "machine-aided cognition" and "multi-access computing," because in the early 1960s computer systems design and AI research had not yet parted ways.

Down in Building 26, where the dirty work went on, where this motley group of exceptionally gifted programmers got their fingers into the logical guts of the machines and made them do their bidding, they were Maniacs And Clowns, Men Against Computers, and numerous unprintable variations. They were the unruly but indispensable hired craftsmen of the projects directed by the likes of McCarthy and Minsky and funded by Licklider—the ones who built the software probes their employers launched into the frontiers of machine intelligence.

At the moment David walked in, a young man named Richard Greenblatt, who lived on the stereotypical hacker diet of soft drinks, candy bars, and Rolaids, and who didn't stop to sleep, much less to wash or change clothing, was explaining to a circle of awed admirers, which included some of the computer scientists who had hired him, how he intended to write a chess playing program good enough to beat a human. Greenblatt's thesis advisor, Marvin Minsky, tried to discourage Greenblatt, telling him that there was little hope of making progress in chess playing software.

Six years after he first stumbled upon the inhabitants of Building 26, sixteen-year-old David Rodman, by now a dropout, acidhead, and professional AI programmer of his own, albeit smaller, repute, was in the group that watched Greenblatt's "MacHack" program demolish Hubert Dreyfus, the number one critic of the whole AI field, in a much-heralded and highly symbolic game of chess. The MacHack versus Dreyfus duel has become one of the hacker legends, and MacHack became the first program to be granted honorary membership in the American Chess Federation.

The Dreyfus chess match was only one of several historic moments in AI history that David witnessed from his vantage point of mascot, then apprentice, then full-fledged hired hacker, during the heyday of MAC, between 1960 and 1967. He was there when his motley colleagues began to build the programming and operating systems for the TX-0 and PDP-1 computer hardware, thus establishing the first software thrust into the age of interactive computing. David was also there when Joseph Weizenbaum, to his later regret, unveiled ELIZA, probably the most widely quoted and widely misunderstood program in history—the program that seems to be an uncannily perceptive psychiatrist, but is actually a programmer's semantic trick.

Project MAC programmers, circa 1959–1960, before hackers looked any differ-ent from other MIT students. (Photo by J. Ph. Charbonnier. Courtesy of the MIT Museum.)

David came upon the hackers through a mixture of mischief and happenstance. He was one of those prodigies who was angry about having a brain like his trapped for another eight years in the body of a child. Since he was six, he had been an exceptional musician, but he gave up the piano at ten because he despised performing for adults. He was a loner, a wanderer, a looker through doorways, an urban spelunker—a snoop, but not a thief, unless you consider knowledge of how to find your way through a complicated system as a stealable property. By the age of fifteen, David and his friends could find their way into any building in the MIT complex, via the system of underground utility tunnels.

Wandering through the halls of MIT, where his father worked in the medical school, was one of his favorite pastimes. He liked to try doors and see what was behind the unlocked ones. When he cast his eyes on those strange guys gathered around an odd-looking television set with wires coming out of it, and then joined them at a game they called "Spacewar," using a control panel made out of a cigar box, and nobody seemed to notice that he was ten years old—David knew he had found his new intellectual home.

"They treated me with some subtlety. I think it was a kind of recognition. They had all been through it, but they weren't about to tell me anything before I figured it out for myself," David recalled, twenty years later. He just sat down and there was a keyboard and someone got him started, and although they were the first people he had met who didn't make a fuss over his intelligence, they noticed how quick he picked it up, all right.

After David returned a few times, and demonstrated his ability to find his way around the computer, the hackers made him a mascot, and when he was a full-blooded initiate ("when they started calling me 'Rodman' instead of 'hey, kid' "), they started giving him small tasks in machine language, eventually showing him tricks in the sexy new programming language known as LISP invented specifically for AI programmers by John McCarthy, one of the project's founders.

Marvin Minsky's secretary took a liking to this wiseass ten-year-old who seemed to take to programming as some kids take to chess or tennis or ballet, and Minsky, who has always been the hackers' patron in MIT computer circles, let David use his password.

Today, having grown up through the early days of the hackers and AI research, the ARPAnet years, the consulting contracts and security clearances, the regular escalation of his income, and the transformation of the social status of computer programmers from weirdo outsiders to millionaire culture heroes, David Rodman is the president of a microcomputer software company whose primary product is a system of programs he wrote himself. His personal odyssey from the inner sanctums of AI hackdom to the rough-and-tumble capitalism of the microcomputer software industry is a kind of capsule history of the whole strange journey of interactive computing from laboratory curiosity to home appliance.

But like many others who are now in their middle thirties and who didn't always wear suits and carry briefcases, his early history was colorful and not a little painful: "At the age of ten, I was like a coiled spring inside—lonely, uptight, angry, cynical. I was unable to balance my intelligence against the rest of the world. Then suddenly, here were people not unlike myself, who showed me a device that would respond to me when I sat down to program it. Those people *knew* what was happening to me, and when I began programming, they encouraged it."

MIT, to begin with, was the engineers' school of engineers' schools, where the undergraduates hold an annual "ugliest man on campus" contest—an unashamed, self-proclaimed, national haven for supernerds. The campus population was primarily composed of the people from all the high schools in the country who stayed home and learned integral calculus or built ham radios while everybody else was at the sock hop.

Amid all this self-styled rejection of conventional youth culture and the atmosphere of cultivated unfashionability, the computer obsessives were considered oddballs even by the other outcasts. Their standards were entirely their own. They and their computers, and a few people in ARPA, were the only ones who knew that the top hackers were really the insiders. Although they were outcasts from the wider society, from their fellow techies, and even from most other computer scientists, they happened to be the people who were creating the future of computing—the first time-sharing systems.

They were having so much fun with what they all knew to be the hot technology of the future that they seemed to deliberately encourage their unappetizing image. You don't just barge in and make yourself a hacker. You've gotta hack. And that means making a computer do things its manufacturers never expected it to do. (This kind of programming is known among hackers as "black magic.") It also meant surviving what the other hackers could do to the results of all your work if you weren't clever enough to prevent them.

There was a matter of intellectual style. Boldness and speed and raw power were as important as (critics of hackers would say *more* important than) elegance and efficiency when it came to "cutting code" (writing the detailed machine language or high-level language lists of instructions that make programs do what computer users want them to do). One common comeback when an outsider asked what "hacker" meant was "somebody who makes furniture with an axe." Orthodox programming style was hardly de rigeur in this crowd. The challenge was to think of a clever way to do something that most normal computer experts would do some other way or not at all. The performance standards were idiosyncratic and subtle, but all-important. These people judged each other by criteria that the rest of the world didn't even understand, and the hackers didn't mind keeping it that way.

They were other kinds of outcasts besides social outcasts, self-selected or otherwise. Their values were entirely their own: academic or commercial success was too trivial to be considered a driving motivation; the opportunity to work with like-minded colleagues on state-of-the-art equipment was paramount. They had their own culture, their own ethic, even their own dialect. The eighteen-year-old MIT dropouts David Rodman wanted to emulate were distinguished from the hippies and radicals they superficially resembled because they all happened to have a talent that was particularly valued in those days, and still is—the ability to write the code that makes computers useful to nonprogrammers.

While all their former classmates were on to their doctorates and assistant professorships and corporate research laboratories, the misfits suddenly found themselves making more money than their conventionally successful peers, at a job where they weren't relegated to working out a payroll system or an airline reservation service. The hackers knew, even if nobody else did, that they—and not IBM, or even their straighter "FORTRAN type" colleagues in computer science—were the test pilots of the computational frontier, pushing the limits of what could be done with each fresh generation of hardware.

Their mandate was to dream up new things for computers to do, and in the process what they did was invent a whole new computer system and computer-oriented subsociety, a technology and social order in which their own little fraternity of ex-outsiders, and not the conventional computer types, were privileged to know the inner mysteries. When the rest of the world caught up with them, they knew they would be on to something even more mysterious to the outsider and more exciting to the hacker. None of them would deny the charges of addiction. Some of the same people who were in that room when David walked in, almost a quarter of a century ago, are still sitting in front of a computer terminal, somewhere on the upper floors of 545 Technology Square.

Their superiors were smart enough to know that the best of the hackers would come up with amazing things if they were left to their own devices. Spacewar, which spread from MIT to other campus computer centers, was one of the rites of passage and defining characteristics of any den of hackers. It was invented by a MAC hacker named Russell, known as "Slug," was perfected in a communal effort by generations of hackers, and it survived wherever it sprouted, like some antibiotic-resistant microorganism, because every computer laboratory manager in the country learned that programmer productivity dove when Spacewar was banned and shot back up when the game was reinstated.

It was Spacewar that influenced Nolan Bushnell to create, over a decade later, a much simpler version called *Pong*, a commercial venture that created the first incarnation of Atari Corporation and a billion-dollar video game industry. Before Pong succeeded, however, Bushnell had failed to get people interested in a more complex game, a more direct derivation of Spacewar. But in those days, the people who put quarters into machines at bars and arcades hadn't yet been educated in their video game sensibilities by the Space Invaders and Pac-Man phenomena of the late 1970s and early 1980s.

But fun and games were only part of the fun and games. One of the things the hackers were building when David arrived was the software for one of the first time-sharing systems. They were writing a time-

Professors Robert Fano and Marvin Minsky, directors of Project MAC, with a PDP-1, lightpen, and very early computer graphics program. (Courtesy of the MIT Museum.)

sharing operating system that they intended to use to create the greatest hacks, the biggest pranks, the most amazing demonstrations of programming virtuosity in hacker history. The fact that they were pioneering a whole new way to use computers that would eventually bring the outside population in on it was not the first thing on their mind. They wanted to get their own hands on the system, so they built it in record time.

Actually, there were two MIT time-sharing projects. The more staid project was CTSS—Compatible Time-Sharing System, so named because it was designed to be compatible with other systems that were being constructed elsewhere. The MAC hackers were designing an operating system they called ITS—the Incompatible Time-sharing System. They couldn't care less about making it easy for outsiders to use. They were having too much fun to share it with the kind of straight-arrow programmers who could stand to eat or sleep before finishing a good hack.

There were hackers and there were metahackers. Richard Greenblatt, because of what his program did to Dreyfus, and because of his ability to improvise great code without fully understanding how he did it, was at the top of the hacking order. He was a dropout and looked the part of a "Pepsi-guzzling, nonsleeping, single-minded programming addict who ate only food that came from a vending machine and whose skin had not absorbed anything but fluorescent light in three years," as Rodman fondly remembered him, three decades later. But Greenblatt's peers knew him as a Nijinsky, a Frank Lloyd Wright, a Johann Sebastian Bach of LISP programming.

The matter of pranks, of what the hackers called "wheel wars"— mucking up each other's files, trying to thwart each other or "crash" the operating system—was part of the working environment. Crashing the system could be accomplished by running some kind of unrunnable self-swallowing program that the programmer who designed the system hadn't made precautions for. When such a prank succeeds, everybody connected to the system at the time can lose important data. In the early sixties, at places like MAC, it was understood that, despite its unfortunate side effects, crashing was an allowable test of the system if the hack revealed an important system vulnerability.

Two decades later, when mischievous and sometimes vandalistic teenagers with home computers started calling themselves "hackers" and crashed the files of nonhackers via the telephone, they were doing something quite different in its ultimate effect, if not in its outward appearance, from what the first such outlaws at MAC were trying to accomplish. The excuse that they were "just exploring" an interesting vulnerability in the system had some real validity back when the hackers were creating and testing new time-sharing systems, and when their expertise was aimed toward a common goal. But when the system that crashes, as nearly happened in 1983, is an operational computer used by a hospital to keep track of patient medication records, it is a somewhat different matter. The same kind of iconoclastic mischief that had one meaning in the 1960s took on another meaning in the 1980s.

"Phone-hacking" was another kind of prank pioneered by MAC hackers in the early 1960s that was to spawn anarchic variants in the 1970s. The self-taught mastery of complex technologies is the hallmark of the hacker's obsession, the conviction that all information (and information delivery technologies) ought to be free is a central tenet of the hacker ethical code, and the global telephone network is a complex technological system *par excellence*, a kind of ad hoc worldwide computer. The fact that a tone generator and a knowledge of switching circuits could provide access to long-distance lines, free of charge, led to a number of legendary phone hacks. But the mythology didn't die there.

In California, the Stanford AI Laboratory (SAIL) and the proximity to Silicon Valley led to the growth of another phone-hacking subcult of "phone phreaks" in the 1970s, whose hero was a fellow who went by the name of Captain Crunch. A gap-toothed, crazy-eyed, full-bearded fellow who now writes software and stays away from illegal activities, Crunch traveled the highways in the late sixties and early seventies with a van full of electronic equipment, playing virtuoso pranks from roadside phone booths—until he was caught, prosecuted, sentenced, and jailed. One of Crunch's phone-hacking buddies from the outlaw days, Steve Wozniak, went on to bigger fame when he invented the first Apple computer. Captain Crunch, also known as John Draper, now makes very decent legitimate money as "Cap'n Software," the sole programmer for the microcomputer software company of the same name.

At Project MAC, and at the subcultural counterparts at Stanford (where they began to blend some of their California brand of craziness into the hacker formula) and elsewhere, you had to suffer in order to be admitted to the more interesting levels of hacker wisdom. As in any other closed subculture, the hackers spared no one their own kind of rites of passage. David was the youngest initiate, but they didn't go any easier on him than any other newcomer. You just weren't part of things at MAC until you met the now-legendary "cookie monster" and some of its nastier relatives.

Crashing the system was a fact of life and an implicit challenge at the higher reaches of hackdom—if you were smart enough to come up with something that the system programmers hadn't guarded against, it was more of an honor than a misdemeanor to bring the computer to a halt, dumping hours or weeks of someone's work. By comparison, the cookie monster was relatively mild. Unlike an operating system crash, the cookie monster struck only selected victims, rather than everybody who was unfortunate enough to be using the system when a crash was perpetrated.

The cookie monster would strike most often at four in the morning. (All-night hacking began with time-sharing systems, not only because it fit in with the hackers' weird self-image, but because time-shared systems run faster at night, when all the nonhackers are out having dates or studying poetry or sleeping or whatever nonhackers do at night in the real world.) You would be looking for a bug somewhere in the two-thousandth line of your program. Suddenly, without warning, the words "I WANT A COOKIE!!" appear on your monitor screen—and all your painstakingly crafted code is relentlessly munched into oblivion by the word COOKIE!!, multiplied over and over until you finally figure it out or (horror of horrors) somebody has to tell you: you have to type in the word COOKIE!! on your keyboard.

In their own way, the MAC hackers were the forerunners of other kinds of psychic desperadoes who appeared on college campuses in the later 1960s. A contempt for middle-class values and an abiding interest in the workings of their own mind were two characteristics that hackers were to share with later subcultures who had nothing to do with computers. David Rodman was a confirmed hacker in the 1960s, when he began to dabble in a very different yet strangely similar outlaw subculture that was springing up in the Cambridge student community.

"I would characterize my first acid trip as a quantum leap into the innards of my own psychology," David recalls today. "Suddenly, there I was—inside myself. I didn't know the path to get in, but there I was. I could observe myself playing the guitar or writing code, and think to myself while improvising, 'Where am I going and how do I know how to go there and what am I really expressing?' It was the trip of all trips."

David thinks that "for my peculiar cognitive style, programming was a perfect preparation for psychedelics, because it allowed me to model a little piece of my personality in the machine, and interact with it. The older hackers would tell me 'never mind what the main program does, we want you to write a program that moves a chess piece on a chessboard,' so I wrote a small, gemlike part of the utility package that went into one of the chess programs. The next time I found myself in one of those gemlike structures was on my first acid trip."

The small "gemlike structures" that David created were incorporated into early versions of Greenblatt's MacHack, the program that eventually became an emblem of the hackers' sovereignty within the AI community when MacHack met Dreyfus in 1967. It all started when Hubert Dreyfus had the temerity to question not only the chances of success but the very legitimacy of AI research. The entire field of artificial intelligence had been challenged as a fraud, and very serious efforts that went beyond the usual acrimony of academic debate were being made to cut off funding for the foolishness Minsky et al. were attempting. The Dreyfus affair began in the summer of 1965, when Hubert Dreyfus—a philosopher, not a computer scientist—spent a few months at the Rand Corporation. The paper Dreyfus wrote at the end of that summer, entitled "Alchemy and Artificial Intelligence," was informally circulated as a Rand report.

Dreyfus thought AI was a crock. He specifically attacked some of the claims AI enthusiasts had made about the future of their field. He proclaimed that the "progress" the AI folks had been citing was an illusion, and attempted to prove that their goal was a delusion. Game playing programs came in for special treatment. An IBM researcher, Arthur Samuels, had recently created a pretty decent checkers program that was on its way to becoming a champion. To Dreyfus, saying that the checkers program represented a step toward a true human-like machine intelligence was like saying that an ape who could climb to the top of a tree was making progress toward flying to the moon.

Dreyfus challenged the idea that a chess playing program of any significance could ever be built, pointed out that in 1957 Herbert Simon had predicted an unbeatable chess playing program within ten years, and noted that the time was about up. Greenblatt came out of nowhere with his carefully constructed chess hack, and Seymour Papert, then codirector of MAC, maneuvered Dreyfus into a public match.

David and the other witnesses remember the game as a dramatic and unpredictable match—a cliff-hanger that was far more suspenseful and ingenious and less mechanical than what any of them had expected. This was more than a friendly rivalry. The source of their funds was being attacked, and it was just possible that this . . . this . . . *philosopher* might manage to get people so stirred up that they would take their precious terminals away. It was a grudge match, no question about it.

MacHack won. Gleefully, the bulletin of the Special Interest Group in Artificial Intelligence (SIGART) of the Association for Computing Machinery reported the results of the match under a headline taken from Dreyfus' paper: "A Ten-Year-Old Can Beat the Machine—Dreyfus." The SIGART editors amended it with a subhead of their own: "But the Machine Can Beat Dreyfus." The SIGART article touched off a series of letters to editors, accusations and counteraccusations, and Dreyfus ended up writing a book, *What Computers Can't Do*, in which he admitted: "Embarrassed by my expose of the disparity between their enthusiasm and their results, AI workers finally produced a reasonably competent program. R. Greenblatt's program called MacHack did in fact beat the author, a rank amateur." [1]

MacHack went on to become an honorary member of the U.S. Chess Federation, and the Dreyfus-versus-AI controversy has dragged on for decades, albeit without the hand-to-hand fury of 1967, when a hacker rose brilliantly to the defense of his art with a legendary hack, then retreated back to his terminal while others argued the significance of what he had done. The event had more than symbolic significance: the formal paper Greenblatt wrote about the program was of historical value to those who still hope to fulfill Turing's, von Neumann's, and Shannon's dreams of playing against a true master chess-machine.[2]

MacHack was actually the second of the two historic software births David Rodman witnessed during his apprenticeship at MAC. Joseph Weizenbaum showed up at MIT in 1963, and when he created ELIZA between 1964 and 1966, he changed the way everybody thought about what computers can't do—and that included changing his own mind about where the whole computer-AI enterprise was heading. ELIZA was a clever way of mimicking human interaction through a computer-mediated dialogue; what the inventor hadn't anticipated was people's willingness to be taken in by the mimicry—even people who should have known better. By the time Weizenbaum recovered from the shock of seeing the way people reacted to his program, he was convinced that something very dangerous lurked in the much-heralded computer revolution.

The reaction to ELIZA eventually led Weizenbaum to question the ultimate value of the changes that computers were introducing to the general population—changes he felt we might all later regret. He also declared that we would soon be faced with important decisions about what computers ought and ought not to do. He specifically cited the hackers as a symptom of a sickness in the heart of computerdom. Weizenbaum's assault on some of the most fundamental premises of the computer culture with the 1976 publication of *Computer Power and Human Reason* set off a continuing, oft-heated public debate between Weizenbaum and the AI community.

The Dreyfus-AI debate had been largely a technical argument, which helped make MacHack's technical victory so sweet. Weizenbaum's was a moral argument, and it carried a passionate force far different in effect from that of Hubert Dreyfus, flying in from California with his phenomenology. This was Joseph Weizenbaum, honored professor of computer science at MIT, saying that AI might not be a crock, but we better be a lot more careful with computers, and watch out for the hackers in the process.

Remember when those funny-looking "computer letters" started appearing on the bottom of checks, in the early 1960s? That was part of Joseph Weizenbaum's handiwork in the days before he came to MIT. As a software expert for General Electric, he was centrally involved in

Bank of America's ERMA project, a milestone in the computerization of the world's banking system. When Weizenbaum later spoke about the morality of using computers in ways that might change millions of people's lives, he was speaking from experience. His creation of a program that gave the illusion of a wise, all-knowing, computerized psychiatrist— and his shock at seeing how willingly even his computer-sophisticated colleagues were taken in by the illusion—triggered Weizenbaum's dissension.

Weizenbaum started out at MIT with what he thought would be an abstract interest in programs that used simple programming tricks to answer questions posed in English. He put together a working version that the hackers had fun with, and which seemed to be a step, albeit a very primitive first step, toward a genuine language-understanding program. The hackers enjoyed the pretense that they were "conversing" with a computer, even though they knew that the program was parsing very simple sentences with no real underlying understanding of their meaning.

While he was working on a more elaborate question-answering program, with greater sentence-parsing capabilities than his first version, Weizenbaum met a psychiatrist from Stanford named Kenneth Colby, who was interested in using computers to model mental states and mental disorders in order to find clues to more effective methods of psychotherapy. Colby helped Weizenbaum create a special-case version of ELIZA, known as DOCTOR. In the introduction of his 1976 book, Weizenbaum described how this most successful and disturbing variation of ELIZA came into being, and conveyed a little of the flavor of the "ELIZA effect": [3]

> The work was done in the period 1964–1966, and it was reported in the computer-science literature in January 1966 and August 1967. To summarize it briefly, I composed a computer program with which one could "converse" in English. The human conversationalist partner would type his portion of the conversation on a typewriter connected to a computer, and the computer, under control of my program, would analyze the message that had so been transmitted to it, compose a response to it in English, and cause the response to be typed on the computer's typewriter.
>
> I chose the name ELIZA for the language analysis program because, like the Eliza of Pygmalion fame, it could be taught to "speak" increasingly well. . . .
>
> For my first experiment, I gave ELIZA a script designed to permit it to play (I should really say parody) the role of a Rogerian psychothera-

pist engaged in an initial interview with a patient. The Rogerian psychotherapist is relatively easy to imitate because much of his technique consists of drawing his patient out by reflecting the patient's statements back to him. The following conversation between a young lady and ELIZA playing doctor illustrates both the Rogerian technique of encouraging a patient to keep talking and the operation of the computer program ELIZA. The first to "speak" is the young lady. The computer's responses are printed entirely in capitals.

> Men are all alike.
> IN WHAT WAY
> They're always bugging us about something or other.
> CAN YOU THINK OF A SPECIFIC EXAMPLE
> Well, my boyfriend made me come here.
> YOUR BOYFRIEND MADE YOU COME HERE
> He says I'm depressed much of the time.
> I'M SORRY TO HEAR YOU ARE DEPRESSED

The first thing that alarmed Weizenbaum was the way people who knew enough about computers to know better began to get drawn into serious conversations with it about their lives! Even his own secretary had fallen into the habit of conversing with it! People were asking to be left alone with the machine to spill out their most intimate thoughts! Weizenbaum was so horrified that he started rethinking everything he believed. But that was just a symptom of how gullible we all might be to what he knew was just a clever hack.

More serious, to Weizenbaum's way of thinking, was the fact that Colby and others thought that "computer-therapists" might soon be applied to some form of "automatic psychotherapy"—an idea Weizenbaum considered "obscene." Weizenbaum's ethical debate, although it originated in the same laboratory that spawned so many important innovations in AI and computer systems design, will not be discussed at length here. His book and the ideas expressed by Weizenbaum and his critics deserve consideration on their own accord.

David Rodman was one of those who spent time conversing with ELIZA when it was still in its infancy, while he was employed as a research assistant in the same laboratory. Some of David's earliest LISP hacks were attempts to emulate ELIZA. And although Weizenbaum didn't know it, some of David's early acid trips were spent in "conversation" with ELIZA.

While Minsky was a kind of patron saint of hackdom, and Greenblatt was an unkempt hero, and McCarthy had his own brand of AI prodigies, Weizenbaum was not very fond of some of the hackers who shared his working quarters, to put it mildly. In his book, he mounted a direct assault on the inner circle of hard-core hackers: [4]

> Wherever computer centers have become established, that is to say, in countless places in the United States, as well as in virtually all other industrial regions of the world, bright young men of disheveled appearance, often with sunken glowing eyes, can be seen sitting at computer consoles, their arms tensed and waiting to fire, their fingers, already poised to strike at the buttons and keys on which their attention seems to be as riveted as a gambler's on the rolling dice. When not so transfixed, they often sit at tables strewn with computer printouts over which they pore like possessed students of a cabalistic text. They work until they nearly drop, twenty, thirty hours at a time. Their food, if they arrange it, is brought to them: coffee, cokes, sandwiches. If possible they sleep on cots near the computer. But only for a few hours—then back to the console or the printouts. Their rumpled clothes, their unwashed and unshaven faces, and their uncombed hair all testify that they are oblivious to their bodies and to the world in which they move. They exist, at least when so engaged, only through and for the computers. These are computer bums, compulsive programmers. They are an international phenomenon.

Weizenbaum attacked those particularly obsessed specimens among the hackers he called "compulsive programmers" on several grounds besides their unorthodox appearance and dietary habits. But he also took care to note (parenthetically) that "(It has to be said that not all hackers are pathologically compulsive programmers. Indeed, were it not for the often, in its own terms, highly creative labor of people who proudly claim the title 'hacker,' few of today's sophisticated time-sharing systems, computer language translators, computer graphics systems, etc., would exist.)" [5]

The compulsive programmers, according to Weizenbaum's criteria, spend far more time playing with their computers than using them to solve the problems they are being paid to solve. They are often superb technicians, he admitted, but he also charged that they are very often so sloppy when they document the programs they have written that other programmers, when they later have to use or modify them, are unable to make sense of what they did.

The obsessed hacker's motivation is not problem-solving, but the raw thrill of interacting with the computer, and that, Weizenbaum charged, was a sign, not of prodigy, but of pathology. "The compulsive programmer," he insisted, "is merely the proverbial mad scientist who has been given a theater, the computer, in which he can, and does, play out his fantasies."

Minsky and others rose to the hackers' defense, pointing out that they should be considered with some of the same suspension of normal standards that society reserves for artists. And just as it is true that a hollow-eyed dropout is not a particularly pleasant sight, and perhaps there is truth to the charge that many of them find it easier to relate to the machine than to other people, isn't there also a chance that they are being unfairly maligned?

Hackers would rather be judged by their creations than by their behavior, and nobody cares about van Gogh's habits of dressing, or whether Mozart went without sleep for days at a time. Minsky deplored the public stereotyping and scapegoating of people who happen to be passionate about programming instead of violin playing or basketball or making money.

Weizenbaum was undoubtedly right about the temptation to use computers for simulating fantasies of omnipotence over fantastically controllable worlds. The value to society of obsessively converting sophisticated computers into toys and games has been a matter of extended debate. Nobody would deny that hackers love fantasy. That these fantasies can be fascinating to nonhackers as well has been an inside secret for years, ever since the hack known as "Dwarfs Hall of Mists, XYZZY and the Infamous Repository," created by Will Crowther and Don Woods, now more commonly known as "Adventure," surfaced at MAC and SAIL.

After they introduce you to ELIZA, "Adventure" is what hackers show you when you ask them why they are addicted to computing. They hit a few keys, sit you down in front of a monitor and a keyboard, and come back in a few hours to forcibly unplug you. Even in this age of more sensually dazzling computer-generated effects, the sheer temptation to explore the computer-stored fantasy remains strong.

After you are told that you can give simple instructions like "drop sword," "go up," "cross bridge," the following words, still famous at every campus computer center, appear on the screen: "You are standing at the end of a road before a small brick building. Around you is a forest. A small stream flows out of the building and down a gulley . . ."

Without warning, and without any high-resolution graphics or sound effects, you are drawn into Colossal Cave, where a labyrinth of chambers containing treasure, dwarfs, magic, strategy and dangers awaits your command. It can take weeks to finish a game. More than one commentator has used "Adventure" as a metaphor for hacking: There is a complex pathway hidden inside the computer, and it is up to the hacker to use all his or her skill, knowledge, and magic to find the treasure and bring it back.

A high regard for programming skill, a mischievous bent, and a predilection for playing games seemed to accompany the spread of hacker culture, along with Spacewar and Adventure. Weizenbaum might have been the first, but he wasn't the last computer scientist to voice concern over the possible dangerous side effects of this way of thinking.

One famous debate erupted at Stanford, years after Weizenbaum's original diatribe. Stanford has been a West Coast headquarters for hackers since the mid-1960s, although significant outposts have long existed at UC Berkeley, Los Angeles, San Diego, and Santa Barbara, at Stanford Research Institute, and even at Rand before the Ellsberg affair. But LOTS—Stanford's Low Overhead Time-sharing System—is where the undergraduate hackers hang out. It was here that another, more recent major hacker controversy surfaced, in the form of a dialogue on the medium that was known by the mid-1970s as "electronic mail." It was the option of everybody on LOTS to post and read messages, either to specific individuals or groups, or to anyone who was interested, via the "bulletin board" sector of the mail program. People could read and add messages whenever they were logged onto the computer.

Sometime serious issues were discussed in this manner, and sometimes long impassioned graffiti (known as "flames") were launched against a variety of targets ranging from the profound to the utterly inconsequential. Sometimes serious issues were disguised as flames, and vice versa. Branches and subbranches of such exchanges could continue for months, making up a kind of electronically embedded ad hoc literature. That was where the "hacker papers" came from.

This particular counterpoint of flames on the subject of hackers, written by hackers, came to the attention of the "real world" because a Stanford professor of psychology named Philip Zimbardo discovered the dialogue and published it, with commentary, in *Psychology Today* magazine in 1980, twenty years after Rodman met Greenblatt et al. in Building 26.

The exchange of flames began with a hacker's version of Luther's 95 theses, nailed, metaphorically, to the door of the electronic temple. A self-sworn ex-hacker who called himself "G. Gandalf" (the tradition is to give oneself a pseudonym on the public mail channel, like the "handles" used in the citizen's-band radio subculture) posted a bulletin entitled "Essay on Hacking," that said, among other things: [6]

> In the middle of Stanford University there is a large concrete-and-glass building filled with computer terminals. When one enters this building through the glass doors, one steps into a different culture. Fifty people stare at terminal screens. Fifty faces connected to 50 bodies connected to 50 sets of fingers that pound on 50 keyboards ultimately linked to a computer. . . . These are the members of a subculture so foreign to most outsiders that it not only walls itself off but is walled off, in turn, by those who cannot understand it. The wall is built from both sides at once.
>
> These people deserve a description. In very few ways do they seem average. First, they are all bright, so bright, in fact, that they experienced social problems even before they became interested in computers. Second, they are self-contained. Their entire social existence usually centers around one another. . . . Third, all aspects of their existence reinforce one another. They go to school in order to learn about computers, they work at jobs in programming and computer maintenance, and they lead their social lives with hackers. Academically, socially, and in the world of cash, computers are the focus of their existence.

As might well be expected, this diatribe did not go unanswered. As usual, opinion was heatedly divided. Some—a minority, of course—agreed wholeheartedly with the heresy. Hackers as a group harbor a love for heresy, iconoclasm, and debating whether something is or is not heretical, even if—especially if—the topic relates to hackers themselves.

Of those who rebutted Gandalf, the one known as "A. Anonymous" offered the West Coast version of the "Minsky defense": [7]

> We are dealing with an infinitely malleable tool. People who choose to develop and use that tool, whether for work, play, or both, have that choice and cannot be denied it. A person who chooses to be a musician must devote hours and hours to gain adequate expertise. But would you consider the computer hacker any less creative than such a person? I certainly wouldn't. The computer serves not only as a workhorse, but also as an easel for exercising one's creative abilities. Therefore, in my opinion, the hacker has not limited himself at all. Rather, he has expanded his intellectual horizon because now he has the infinite tool.
>
> As for the charge that it disrupts one's social life, I would tend

to agree with this to a point. But it depends on how controlled the individual is. At any time, he can withdraw to a more normal schedule. Why doesn't he? The reason is obvious. The infinite tool that knows few boundaries is accessible to a much higher degree, and thus he can devote more time to it. Why is this wrong? I think it is definitely a bonus, since the usual restraints of 9-to-5 are eliminated and the person is allowed to expand beyond boundaries to do what he wants.

Now we come to the human versus the machine factor. Gandalf stresses the necessity of human interaction and the inherent evil of the machine. Would you stress the evil of the instruments in an orchestra, or the instruments in a laboratory, or the typewriter of an author? All of these occupations demand extraordinary amounts of time for excellence. But I see no greater human interaction in these fields than in computers. I feel that people who disparage computers for a seemingly decreased human interaction are not at all familiar with the true import of the computer. Not only is it the infinite tool, it is also an extremely fluid medium of communication.

The publication of the controversy set off an avalanche of electronic mail over the ARPAnet and at local computer centers. The hacker debates had spread to the amateur "bulletin board computers" by 1983, when the movie *WarGames* and the real-life young computer-systems "crackers" who subsequently surfaced brought the word *hacker* to widespread public attention, in this newer, unpleasantly restricted sense.

One of the oldest rules of the game is "thou shalt not do unto ordinary computer users what thou hast done to other hackers." Almost all of the old-time hackers deplore what the young computer trespassers and system crashers did—"dark-side hacking"—although the anarchist minority still insist that the ultimate freedom is the freedom to figure out how the communication-computing system works, and declare that the burden of protection against trespassing ought to be on the system programmer who has files to protect, not on the explorer who might tap in during some midnight jaunt through the network.

Real computer criminals aside, the concern of the noncomputing public over the hacker controversy does seem a bit strange. After all, these people aren't accused of mayhem or arson—just of being very smart when it comes to knowing how to operate computers. The capacity for scapegoating is very high in a culture where most people have been led to believe that computers are either smarter than they are or too complicated for ordinary people to use. James Milojkovic, an associate of Zimbardo's at Stanford who was writing his psychology doctoral thesis about the cognitive and motivational impact of the microcomputer, came to the hackers' defense.

In a 1982 interview, Milojkovic said he spent plenty of time around hackers, and saw nothing pathological about what they are doing. In regard to all the public concern about what threat (noncriminal) hackers might pose, he said: "Clearly it's nonsense. I think what's happening is that there's some sort of fear that maybe what they're doing with the machines is aimed against us." Like "A. Anonymous," Milojkovic sees nothing wrong with a little compulsiveness in regard to learning: "I can think of nothing more natural than to fall in love with knowledge," he said, "and hackers are so deeply in love with knowledge of the computer that they're just swept off their feet." [8]

A case in point: David Rodman. When last we saw him, lurking in the background of the MacHack versus Dreyfus match, an acidhead teenage dropout hacker, he was almost certainly headed for a sunken-eyed, computer-nut future. In fact, quite the opposite turned out to be the case. He was doing quite well for himself, even at sixteen, as a freelance programmer. He got some offers to set up computer systems for social service bureaucrats, so he moved to D.C. in his early twenties.

By 1972, David found himself up to his ears in the same problem that plagued Herman Hollerith—handling huge data bases. In fact, designing probes of the U.S. Census information, now stored on magnetic tape, was David's specialty. He was an expert in extracting information from very large masses of data without spending too much money doing it. He moved back to Cambridge to work for a software think tank, did more than a few jobs for agencies he doesn't want to name, and in 1978 he decided it was time to turn what he knew into a marketable product.

David Rodman ended up creating and marketing a tool for managing data bases, a program that he designed to be usable by microcomputer owners. Thus he was one of many formerly sequestered programmers who joined the software business at the beginning of the consumer computing boom, when it was still possible for a programmer-turned-entrepreneur to go far and fast. A couple of other, older, MIT hackers put out VisiCalc in 1978—the "electronic spreadsheet" that allows users to ask "what-if" questions about numerical data—and millions of people who had never touched a keyboard before began tackling problems that had formerly been reserved for mainframe programmers.

I first met David Rodman in the early 1980s, because of his strange grin. I knew his name because it was stamped onto the plastic card that was pinned to his lapel. His rumpled suit and convention badge didn't exactly mark him as a high roller, but his smile projected a self-assurance of almost demented intensity. We were standing in the magnificent casino that is conveniently located between the Hilton lobby and the indoor walkway to the Las Vegas convention center. Upward of fifty thousand people attending Comdex, a national convention for the microcomputer industry, trooped through the casino every day. The arriving computerists didn't mind spending their money, and they were an amiable group. A lot of them seemed downright *happy*. David Rodman, for example, was still smiling after he turned away from the craps table.

"Why do you look so damn cheerful?" I couldn't help inquiring.

"I was wrong about the dice," he replied, "but I'm too far ahead to complain."

"Craps?"

"Data base management systems."

"Not my game," I said. "What's the product?"

"About forty pages of zeroes and ones."

"The market pretty good for zeroes and ones?"

"The software market, as of today, is nothing less than astounding."

Considering the fact that he had just dropped a hundred dollars in less than ten seconds, he must have been doing very well indeed to be making money at the convention faster than he was losing it in the casino. The crooked grin on his face, a variant of the slightly demented expression that attracted my attention back at the craps table, made it clear that he didn't mind talking about his business.

We got to know each other, and eventually I learned about what he did before he was the prime mover and chief asset of a software corporation. There was no outward sign that here was an ex-MAC hacker, ex-acidhead, ex-consultant to unnamed intelligence agencies. He was freckled, balding, and what hair he had left was short and neatly combed. He was clean-shaven, and his attire wouldn't have been out of place on an accountant or a widget salesman. But in his heart, he was still a hacker, and an evangelistic one at that.

By the time we got through the story to the point of talking about his current product, it was clear that he had not turned his back on the programming priesthood, but was merely interested in expanding it, to his own profit, by giving millions of people a direct taste of the same experience that hooked him back in Building 26.

"I remember the way I learned jazz improvisation, and how that affected my programming. When I was first learning, I said to myself,

'Here I am in this chord, and I've got to get to that chord.' The transition, the way you hop from note to note or pass a variable from procedure to procedure—that's where the individual style of the musician or the programmer comes in. Nothing happened, a lot of the time. But when my teacher showed me something I hadn't realized before, pointed out that a certain note would work in a way I wasn't expecting, for instance, I would get a little shock of understanding, and the next time I came to a transition I'd loosen my grip on my conscious effort and try to recapture that shock, and there would be the note or the line of code I needed.

"Now think of the person sitting in front of his computer with a keyboard. What this person needs is a profit and loss statement, or information about sales accounts, or a breakdown of stock in inventory. What I need to do is to create an environment for that person, structured in such a way that it is natural and easy to translate his or her desire to the actual P & L statement, or a sales report or inventory account, and even show them how they can improvise along the way. Not only should this tool work better than their old pencil and paper and calculator and filing cabinet—it should also give the user one of those little pleasurable shocks. I want my file management system to enable that person to become a jazz musician.

"A really good program designer makes an artist out of the person who uses the computer, by creating a world that puts them in the position of 'Here's the keyboard, and here's the screen. Now once you learn a few rudimentary communication skills, you can be a superstar.' "

It was an unexpected, but perhaps not inappropriate philosophy to hear from a LISP hacker turned software vendor. He has yet to carve out an empire like Bill Gates or Steve Wozniak, but David Rodman knows that most of the potential consumers of microcomputer software are still in the earliest stages of their progression toward obsessive software intoxication. David sees a niche for people like himself as toolmakers and trailblazers, leading the way for the emergence of an entire population of programming artists. He wants programming to become a performing art.

But long before hackers started thinking about using their computers for intellectual improvisation—before David Rodman was born, in fact— a dreamer out in California was designing his own kind of mind amplifier.

CHAPTER NINE

The Loneliness of a Long-Distance Thinker

Harry Truman was President and *Sputnik* was a word that only Russian language experts knew when Doug Engelbart first thought about displaying words and images on radar screens, storing them in computers, and manipulating them with levers and buttons and keyboards. For over thirty years, Engelbart has been trying to hasten what he believes will be the biggest step in cultural evolution since the printing press. To hear him tell it today, both the computer establishment and the computer revolutionaries still fail to understand that the art and power of using a computer as a mind amplifier are not in how the amplifier works but in what amplified minds are able to accomplish.

At the end of the summer of 1945, just after the surrender of Japan, Engelbart was a twenty-year-old American naval radar technician, waiting for his ship home from the Philippines. One muggy day, he wandered into a Red Cross library that was built up on stilts, like a native hut.

"It was quiet and cool and airy inside, with lots of polished bamboo and books. That was where I ran across that article by Vannevar Bush," Engelbart recalls. More than three decades later, he still fondly remembers the room where he first encountered the dream that has dominated most of his life. At that time, the news of Hiroshima was still fresh

174

and searing. He found himself wondering whether the same inventiveness that produced nuclear bombs might be used to prevent such destruction in the future. Engelbart started designing computer-based problem-solving systems in 1951. He hasn't stopped yet.

The earliest and one of the clearest articulations of the idea that information processing technology could be used to amplify human memory and thinking was the one Doug found that day in 1945, in an article entitled "As We May Think," published toward the end of the war in *The Atlantic Monthly*. The author was the highest-ranking scientific administrator in the U.S. war effort, Vannevar Bush.

Bush, the son and grandson of Yankee seafarers, was the same mathematician who had constructed analog computers at MIT in the 1930s. He was also in charge of over 6000 U.S. scientists during World War II, as director of the Office of Research and Development. His two most important goals were starting the Manhattan Project and finding a means to stop German bombing, goals that both directly hastened the invention of computing machinery. Ironically, Bush didn't mention the potential of the early computers as information-handling devices when he wrote his article. But he did present an idea that was to bear fruit many years later—a description of a science-fiction-like general-purpose tool to help us keep track of what we know.

Looking toward the postwar world, Bush foresaw that recent breakthroughs in science and technology were going to create problems of their own. With all these scientists producing all this new knowledge at an unprecedented rate, how was anyone to keep track of it all? How would this rapidly expanding body of knowledge benefit anybody if nobody knew how to get the information they needed?

"The summation of human experience is being expanded at a prodigious rate, and the means we use for threading through the consequent maze to the momentarily important item is the same as was used in the days of square-rigged ships," Bush wrote.[1] He urged men of science to turn their efforts to making the increasingly unwieldy accumulation of human knowledge more accessible to individuals.

But the future technology that Bush foresaw extended beyond the borders of science to the general citizenry. The day was coming when not only scientists but ordinary citizens would be required to navigate through ever-more complicated realms of information. In the pages of the *Atlantic*, Bush proposed that a certain type of device should be developed, a device to improve the quality of human thinking. Because one of its functions was to extend human memory, Bush called his hypothetical machine a *memex*. But Bush was one of the first to see that rapid access to large collections of information could serve as much more than a simple extension of memory. Although he described it in terms of the primitive information technologies of the 1940s, the memex was functionally similar to what is now known as a personal computer— and more.

Some ideas are like seeds. Or viruses. If they are in the air at the right time, they will infect exactly those people who are most susceptible to putting their lives in the idea's service. The notion of a knowledge-extending technology was one of those ideas. Fifteen years after Bush published his *Atlantic* article, J.C.R. Licklider published his article about making computers into a communication medium. But only five years after Bush's article, Doug Engelbart, infected by the idea of creating a mind-extending tool, incubated his own ideas about how to use machines to augment human intelligence.

After the war, with an electrical engineering degree and his experience with radar, Engelbart found a job at Ames Laboratory in California, working on contracts for one of NASA's ancestors, the National Advisory Committee on Aeronautics. After a couple of years at Ames, he asked a woman he had met there to marry him.

"The Monday after we got engaged," Engelbart remembers today, "I was driving to work when I was hit with the shocking realization that *I no longer had any goals*. As a kid who had grown up in the depression, I was imbued with three goals—get an education, get a steady job, get married. Now I had achieved them. Nothing was left."

Doug Engelbart tends to think seriously about things when he finds something worth thinking about. And his own life is certainly not exempt from being an object of his serious thinking. While he drove along a two-lane paved road that is now a freeway, he reckoned he had about five and a half million working minutes remaining in his life. What value did he really want from that investment? At the age of twenty-five, in December of 1950, he started to think about what new goals he might set for himself.

"I dismissed money as a goal fairly early in the decision process. The way I grew up, if you had enough money to get by, that was okay; I never knew anybody who was rich. But by 1950, it looked to me like the world was changing so fast, and our problems were getting so much bigger, that I decided to look for a goal in life that would have the most payoff to mankind."

For several months after he made his decision to commit himself to an appropriately humanitarian enterprise, Doug searched for the right one. He contemplated his situation and skills and thought about the various kinds of crusades he might join. With his radar training, and what he was beginning to learn about computers, Engelbart was also looking for a cause that wouldn't require him to retread his engineering education, or move too far away from his new home. He had a challenging job and a pleasant drive to work. Santa Clara Valley was still the world's largest prune orchard, and the electronics industry had only recently moved out of a couple of garages in Palo Alto. The drive gave him time to think.

Ultimately, the kinds of crusades that appealed to him still didn't satisfy his needs: There just weren't clear-cut ways of organizing one's thoughts to run a crusade. He was an engineer, not a political organizer, and the world was becoming too complicated for anything but the most well-organized crusades. Suddenly, Doug recognized that he was running into the same fundamental issue over and over again.

Engelbart realized, as had Vannevar Bush, that humankind was moving into an era in which the complexity and urgency of global problems were surpassing civilization's time-honored tools for dealing with problems. He also began to understand, as did Licklider a few years later, that handling the informational by-products of problem-solving had itself become the key to all the other problems. The most important task no longer lay in devising new ways to expand our accumulation of knowledge, but in knowing where to look for the answers that were already stored somewhere. "If you can improve our capability to deal with complicated problems, you've made a significant impact on helping humankind. That was the kind of payoff I wanted, so that's what I set out to do."

Although many of the details took decades to work out, the main elements of what he wanted to achieve came to him all at once: "When I first heard about computers, I understood, from my radar experience, that if these machines can show you information on punchcards and printouts on paper, they could write or draw that information on a screen. When I saw the connection between a cathode-ray screen, an information processor, and a medium for representing symbols to a person, it all tumbled together in about a half an hour.

"I started sketching a system in which computers draw symbols on the screen for you, and you can steer it through different information domains with knobs and levers and transducers. I was designing all kinds of things you might want to do if you had a system like the one Vannevar Bush had suggested—how to expand it to a theater-like environment, for example, where you could sit with colleagues and exchange information. God! Think of how that would let you cut loose in solving problems!"

After thirty often-frustrating years of pursuing a dream that the computer industry has long ignored, Doug Engelbart still can't keep the excitement out of his soft voice and the faraway look out of his eyes when he talks about the prospects he foresaw at twenty-five, and has pursued ever since. But he's not sure whether today's generation of computerists, with all their fancy hardware, are getting any closer to the real issues.

Although history has proved him to be an accurate visionary in many ways, but perhaps a less-than-ideal manager of projects and people, and although even his friends use the word "stubborn" in describing his attitudes about his theories, Doug Engelbart still wields the power of a quiet person. The magnetism of his long-envisioned goal is still strong for him, so strong that a good deal of it still radiates when he talks about it. In 1971, his friend Nilo Lindgren described him in *Innovation* magazine: [2]

> When he smiles, his face is wistful and boyish, but once the energy of his forward motion is halted and he stops to ponder, his pale blue eyes seem to express sadness and loneliness. Doug Engelbart's voice, as he greets you, is low and soft, as though muted from having traveled a long distance, as though his words have been attenuated by layers of meditation. There is something diffident yet warm about the man, something gentle yet stubborn in his nature that wins respect.

"He reminds me of Moses parting the Red Sea," is the way Alan Kay describes Engelbart's gentle charisma. Of course, the original Moses never set foot in the Promised Land. And he never had the reputation of being an easy man to work with.

In 1951, Engelbart quit his job at Ames and went to graduate school at the University of California at Berkeley, where one of the first von Neumann architecture computers was being built. That was when he began to notice that not only didn't people know what he was talking about, but some presumably "objective" scientists were overtly hostile. He started saying the wrong things to people who could affect his career, things that simply sounded strange to the other electrical engineers.

"When we get the computer built," this young engineer kept asking, "would it be okay if I use it to *teach* people? Could I hook it up to a keyboard and get a person to interact with the computer? Maybe teach the person typing?" The psychology people thought it was great, but computers were hardly their department. The engineering people said, "There's no way that kind of idea is going to fly."

The interactive stuff was so wild that the people who knew about computers didn't want to hear about it. Back then, you didn't interact with a computer, even if you were a programmer. You gave it your question, in the form of a box of punched cards, and if you had worked very hard at stating the question correctly, you got your answer. Computers weren't meant for direct interaction. And this idea of using them to help people *learn* was downright blasphemy.

After he got his doctorate, Engelbart came to another one of those internally triggered decision points in his life that his dream continued to bring his way. Nobody in his department wanted to listen to talk of building a better way to solve complex problems, and he felt that he would have to construct a whole new academic discipline before he could begin the research he really wanted to do. The university, Engelbart decided, was a place to get his journeyman's card, but not a place to follow his vision.

Thus, young Doctor Engelbart went to the commercial world, looking for an opportunity to develop electronic systems that would eventually help him do what he wanted in terms of augmenting human intellect, and would pay his room and board as he contributed to the development of marketable devices as well. Engelbart brought some of his ideas to a progressive young company down the road in Palo Alto. For a change, here were some people looking to the future. Not too much more than a decade out of electrical engineering school themselves, Bill Hewlett, David Packard, and Barney Oliver (their head of research and development) were enthusiastic about Doug's proposal. A deal was offered. Engelbart drove home, elated. On the way home, in typical Engelbart fashion, Doug started thinking about it.

"I pulled the car over to the first phone booth and called Barney Oliver and said that I just wanted to check my assumption that they saw a future in digital technology and computers—which I thought was a natural path for their electronic instrument company to follow. I had assumed that they knew that the ideas I proposed to them that afternoon were only a bridge to digital electronics. And Barney replied that no, they didn't have any plans for getting into computers. So I said, 'Well, that's a shame, because I guess it cools the deal. I have to go the digital route to pursue the rest of what I want to do.' "

"So my deal with Hewlett-Packard was called off," Doug says, wrapping up the reminiscence with one of his famous wry smiles, adding: "The last time I looked they were number five in the world in computers."

Doug kept looking for the right institutional base. In October, 1957, the very month of Sputnik, he received an offer from an organization in Menlo Park, "across the creek" from Palo Alto, then known as the

Stanford Research Institute. They were interested in conducting research into scientific, military, and commercial applications of computers. One of the people who interviewed him for the SRI job had been a year or two ahead of Doug in the Ph.D. program at Berkeley, and Doug told him about his ideas of getting computers to interact with people, in order to augment their intellect.

"How many people have you already told about that?" he asked Doug.

"None, you're the first one I've told," said Doug.

"Good. Now *don't* tell anyone else. It sounds too crazy. It will prejudice people against you."

So Doug kept quiet about it. For about a year and a half, he earned his living and learned the ropes of the think-tank business and thought about putting his ideas into a written proposal. Then he told his superiors that he was willing to work hard to pay his way at the institute but he really had to have a framework to cultivate his idea—an augmentation laboratory where people and machines could experiment with new ways of creating and sharing knowledge, or at least a project to describe exactly what an augmentation laboratory might be. There was some friction, but eventually he was given the go-ahead.

The U.S. Air Force Office of Scientific Research, ever vigilant for new knowledge about how humans operate machines, provided a small grant. Doug finally had what he wanted—the freedom to explore a field in which he still had no colleagues. "It was lonely work, not having anybody to bounce these ideas off, but finally I got it written down in a paper I finished in 1962 and published in 1963."

Total silence from the computer science community greeted the announcement of the conceptual framework Engelbart had thought about and worked to articulate for over a decade. But the few people who were listening happened to be the right people. Bob Taylor, a young fellow at NASA who was one of the bright technological vanguard of the post-Sputnik era, one of the new breed of research funders who didn't fear innovation as a matter of reflex, pushed some of the earliest funding of Doug's project.

Fortunately, by that time another one of the few people who were able to understand Engelbart's vision, J.C.R. Licklider, was moving ahead with his ARPA funding blitz. As result of Licklider's support, time-sharing was coming along rapidly. By the early sixties, some of the low-level hardware and software tools to build Doug's dreamed-of high-level methodological and conceptual structures were being tested. Licklider and Taylor thought Engelbart was just the kind of forward-thinking researcher they wanted to recruit for the task of finding new and powerful uses for the computational tools their research teams were creating. They were particularly interested in the same paper of Doug's that the mainstream of computer science had chosen to ignore.

The paper that attracted the attention of ARPA and met such a thundering silence from the wider community of computer theorists in 1963 was entitled "A Conceptual Framework for the Augmentation of Man's Intellect." In its introduction, Engelbart presented the manifesto by which he meant to launch an entire new field of human knowledge: [3]

> By "augmenting man's intellect" we mean increasing the capability of a man to approach a complex problem situation, gain comprehension to suit his particular needs, and to derive solutions to problems. Increased capability in this respect is taken to mean a mixture of the following: that comprehension can be gained more quickly; that better comprehension can be gained; that a useful degree of comprehension can be gained where previously the situation was too complex; that solutions can be produced more quickly; that better solutions can be produced; that solutions can be found where previously the human could find none. And by "complex situations" we include the professional problems of diplomats, executives, social scientists, life scientists, physical scientists, attorneys, designers—whether the problem situation exists for twenty minutes or twenty years. We do not speak of isolated clever tricks that help in particular situations. We refer to a way of life in an integrated domain where hunches, cut-and-try, intangibles, and the human "feel for a situation" usefully coexist with powerful concepts, streamlined terminology and notation, sophisticated methods, and high-powered electronic aids.

It was no accident that "hunches, cut-and-try, intangibles," were listed early and "high-powered electronic aids" was listed last. Although he knew that widespread access to digital computers was the only means by which our society could make use of an augmented knowledge system, Engelbart also understood that the hardware was a low-level component of the total system he meant to augment. Human intellect *uses* tools, but the power of the human mind is not itself limited to the tools the human brain automatically provides.

Our culture has given us sophisticated procedures for dealing with problems, procedures that augment our innate capacity for learning new things by giving us the benefit of what others before us have learned. These ways of doing things are the software that creates civilization. A member of a preliterate culture of the remote New Guinea highlands, for example, possesses the same innate mental capabilities as a Western city-dweller, but something else must be added to the repertoire of what that New Guinea highlander knows how to do before he can drive a car, check out a book from a library, or write a letter.

The "something extra," Engelbart emphasized, is not a property of the tool. It isn't the nervous system of the individual that separates the "civilized" person from the "primitive." To certain cultures that we deem primitive, the most sophisticated urbanite is decidedly lacking in necessary survival skills. If the cultural situation of the previous paragraph were reversed, the same ignorance on the part of the displaced person would be evident: If you drop a lifelong New Yorker into the New Guinea highlands, don't expect him or her to know how to build a grass shelter or what to do in a tropical storm. Somebody who knows what to do in those situations has to teach survival skills to the newcomer, thus augmenting his or her innate capacities. It is here that the original augmentation of human intellect comes in—the tools and procedures that cultures make available to individuals: [4]

> Our culture has evolved means for us to organize and utilize our basic capabilities so that we can comprehend truly complex situations and accomplish the processes of devising and implementing problem solutions. The ways in which human capabilities are thus extended are here called *augmentation means*, and we define four basic classes of them:
>
> 1. *Artifacts*—physical objects designed to provide for human comfort, the manipulation of things or materials, and the manipulation of symbols.
> 2. *Language*—the way in which the individual classifies the picture of his world into the concepts that his mind uses to model that world, and the symbols that he attaches to those concepts and uses in consciously manipulating the concepts ("thinking").
> 3. *Methodology*—the methods, procedures, and strategies with which an individual organizes his *goal-centered* (problem-solving) activity.
> 4. *Training*—the conditioning needed by the individual to bring his skills in using augmentation means 1, 2, and 3 to the point where they are operationally effective.
>
> The system we wish to improve can thus be visualized as comprising a trained human being together with his artifacts, language, and methodology. The explicit new system we contemplate will involve as artifacts computers and computer-controlled information-storage, information-handling, and information-display devices. The aspects of the conceptual framework that are discussed here are primarily those relating to the individual's ability to make significant use of such equipment in an integrated system.

The biggest difference between the citizen of preliterate culture and the industrial-world dweller who can perform long division or dial a telephone is not in the brain's "hardware"—the nervous system of the highlander or the urbanite—but in the thinking tools given by the culture. Reading, writing, surviving in a jungle or a city, are examples of culturally transmitted human software. The hypothetical transplanted native, Engelbart points out, can move step by step through an organized program by which he or she may *learn* to drive a car or check out a book from a library.

How do we adapt to new ways of thinking? Engelbart used the metaphor of a *toolkit*, and proposed that we organize our intellectual problem-solving tools in a hierarchy: [5]

> It is likely that each individual develops a certain repertory of process capabilities from which he selects and adapts those that will compose the processes that he executes. This repertory is like a toolkit. Just as the mechanic must know what his tools can do and how to use them, so the intellectual worker must know the capabilities of his tools and have suitable methods, strategies, and rules of thumb for making use of them. All of the process capabilities in the individual's repertory rest ultimately on basic capabilities within him or his artifacts, and the entire repertory represents an integrated, hierarchical structure (which we often call the *repertory hierarchy*).

As an example, Engelbart offered the process of issuing a memorandum—a task that involves putting specific information in a formal package and distributing it to other people. The reason for writing the memo, the memowriter's role in the organization, the intended audience, the importance of the subject matter of the memo to the organization's goals—these are the higher level components of the hierarchy.

At an intermediate level are the skills of marshaling facts, soliciting opinions, thinking, formulating ideas, weighing alternatives, forecasting, making judgments, that go into framing the memo, and all the communication skills that go into putting the memo into form. Toward the bottom of the hierarchy are the artifacts used to prepare the memo and the medium by which it is communicated—typewriter, pencil, paper, interoffice mail.

Engelbart proposed a hypothetical method for boosting the effectiveness of the whole system by introducing an innovative technology into a relatively low level of the hierarchy. "Suppose you had a new writing machine," he wrote, "a high-speed electric typewriter with some very special features." In a few words, he proceeded to describe what is known today as a "word processor."

What might be the effect of such a machine on the memo-writing process? Engelbart's 1963 speculations sound like advertising copy for word processing systems of the 1980s—and more: [6]

> This hypothetical writing machine thus permits you to use a new process of composing text. For instance, trial drafts can rapidly be composed from rearranged excerpts of old drafts, together with new words or passages which you insert by hand typing. Your first draft may represent a free outpouring of thoughts in any order, with the inspection of foregoing thoughts continuously stimulating new considerations and ideas to be entered. If the tangle of thoughts represented by the draft becomes too complex, you can compile a reordered draft quickly. It would be practical for you to accommodate more complexity in the trails of thought you might build in search of the path that suits your needs.
>
> You can integrate your new ideas more easily, and thus harness your creativity more continuously, if you can quickly and flexibly change your working record. If it is easier to update any part of your working record to accommodate new developments in thought or circumstance, you will find it easier to incorporate more complex procedures in your way of doing things. . . .
>
> The important thing to appreciate here is that a direct new innovation in one particular capability can have far-reaching effects throughout the rest of your capability hierarchy. A change can propagate *up* through the capability hierarchy, higher-order capabilities that can utilize the initially changed capability can now reorganize to take special advantage of this change and of the intermediate higher-capability changes. A change can propagate *down* through the hierarchy as a result of new capabilities at the high level and modification possibilities latent in lower levels. These latent capabilities may have been previously unusable in the hierarchy and become usable because of the new capability at the higher level.

While Engelbart was, in fact, suggesting that computers could be used to automate a low-level task like typewriting, the point he wanted to make had to do with the changes in the overall system—the capabilities such an artifact would open up for *thinking* in a more effective, wider-ranging, more articulate, quicker, better-informed manner. That was why he distinguished his proposed new category of computer applications by using the term *augmentation* rather than the more widespread word *automation*.

From Engelbart's point of view, the fact that it took over fifteen more years for word processing to catch on was not as important as the fact that people continue to myopically concentrate on the low-level automation and ignore the more important leverage it makes possible at higher levels. The hypothesis he presented in the 1963 framework was that computers represent a new stage in the evolution of human intellectual capabilities. The *concept manipulation* stage was the earliest, based in biological capabilities of the brain, followed by the stage of *symbol manipulation* based on speech and writing, and the stage of *manual external symbol manipulation*, based on printing.

The computer-based typewriter was an example of the coming fourth stage of *automated external symbol manipulation*, to be brought about by, but not limited to, the application of computers to the process of thinking and communicating: [7]

> In this stage, the symbols with which the human represents the concepts he is manipulating can be arranged before his eyes, moved, stored, recalled, operated upon according to extremely complex rules— all in very rapid response to a minimum amount of information supplied by the human, by means of special cooperative technological devices. In the limit of what we might now imagine, this could be a computer, with which individuals could communicate rapidly and easily, coupled to a three-dimensional color display within which *extremely sophisticated images* could be constructed, the computer being able to execute a wide variety of processes on parts or all of these images in automatic response to human direction. The displays and processes could provide helpful services and could involve concepts not hitherto imagined (e.g., the pregraphic thinker would have been unable to predict the bar graph, the process of long division, or card file systems).
>
> . . . we might imagine some relatively straightforward means of increasing our external symbol-manipulation capability and try to picture the consequent changes that could evolve in our language and methods of thinking. For instance, imagine that our budding technology of a few generations ago had developed an artifact that was essentially a high-speed, semiautomatic table-lookup device, cheap enough for almost everyone to afford and small and light enough to be carried on the person. Assume that individual cartridges sold by manufacturers (publishers) contained the lookup information, that one cartridge could hold the equivalent of an unabridged dictionary, and that a one-paragraph definition could always be located and displayed on the face of the device by the average practiced individual in less than three seconds. What changes in language and methodology might not result?

If it were so easy to look things up, how would our vocabulary develop, how would our habits of exploring the intellectual domains of others shift, how might the sophistication of practical organization mature (if each person could so quickly and easily look up applicable rules), how would our education system change to take advantage of this new external symbol-manipulation capability of students and teachers and administrators?

At the end of the 1963 paper, Engelbart proposed that the hypothesis should be tested by constructing an augmentation laboratory in which humans could use new information processing artifacts to explore the new languages, methods, and training made possible by the computer systems then coming into existence in Cambridge, Lexington, Berkeley, and Santa Monica. Since the ultimate product was to be for everyone, not just computer experts, people who were involved in editing, designing, and other knowledge-related fields would have to be recruited to join the electrical engineers and programmers. Because the goal was to enhance the power of the human mind, and to learn how to introduce such enhancements to human organizations, a psychologist would also be needed.

The laboratory itself would have to be a consciously designed bootstrapping tool, because the very tools this team would be constructing first were the tools they needed to do their own job better. Before they could hope to augment other people's tasks, they had to augment their own jobs. Bootstrapping—building the tools to build better tools, and testing them on yourself as you go along—was a central component of Engelbart's strategy, intended to match the pace of anticipated developments in computer technology. SRI management had few illusions about obtaining the funding necessary to implement such a scheme.

In 1964, Bob Taylor, who by that time had moved from NASA to ARPA, told Engelbart and SRI that the Information Processing Techniques Office was prepared to contribute a million dollars initially to provide one of the new time-sharing computer systems, and about a half a million dollars a year to support the augmentation research. It came as a surprise to Engelbart's superiors, who were eager to procure government contracts for developing new computer technologies, but who didn't exactly regard his grandiose plans for a mind-extending laboratory as their most promising candidate for large-scale funding. One can imagine the SRI brass pulling out the organization chart after the ARPA funders left, to find out who and where Doug Engelbart happened to be.

Here was the support Engelbart had been seeking for years, coming right at the point when the conceptual framework for the system had already been worked out and the technology he needed was becoming available. The next step was to assemble the team who would build the first prototype.

Perhaps the Augmentation Research Center's greatest effect on computer culture for generations to come was in the succession of remarkable people who passed through that laboratory and on to other notable research projects. Dozens of gifted individuals over the span of a decade dedicated themselves to putting into action the system Engelbart and Licklider had dreamed about in previous years. Many of those former Engelbart protégés are now leaders of their own research teams at universities or the R & D divisions of commercial computer manufacturers.

The Augmentation Research Center (ARC) consisted of the "engine room," where the new time-sharing computers were located, a hardware shop where the constantly upgraded computer systems and experimental input-output devices were built and maintained, and a model "intellectual workshop" that consisted of an amphitheater-like space in which a dozen people sat in front of large display terminals, creating the system's software, communicating with each other, and navigating through dimensions of information by means of what was known as *NLS* (for oNLine System).

NLS was an exotic and intoxicating new brew of ARPA-provided gadgetry, homebrewed software wizardry, and altogether new intellectual skills that were partially designed in advance and partially thrown together as the designer-subjects of the experiment went along. After four years of stumbling, backtracking, leaping forward, then more confidently exploring this new territory, after hardware crises and software crises and endless argumentation about how to go about doing what they all agreed ought to be done, NLS was beginning to fulfill the hopes its builders had for it. It was time to gamble.

Whenever he consulted the feeling in his stomach, Doug Engelbart had no doubt that it was a gamble. Sitting all alone on that stage in San Francisco, watching his support team scramble around the hastily woven nest of cables and cameras surrounding the base of the platform, facing an audience of several thousand computer experts, it was all too evident to Doug that any one of a number of possible accidents—a thunderstorm, a faulty cable, a concatenation of software glitches—could effectively kill their future chances of obtaining research funds.

But he had begun to lose his patience, waiting for decades for the rest of the world to catch on to something as important as augmentation. And his colleagues shared Engelbart's confidence in the delicate coalition of people, electronic devices, software, and ideas they called the NLS system.

Doug's painstakingly thought-out conceptual framework, the prototype hardware systems he and Bill English developed, and his bootstrapping laboratory of systems programmers, computer engineers, psychologists, and media specialists were only corroborating what Doug had known for years—computers can help intellectual workers *think* better. By the late 1960s, the problem lay in getting his ideas and the meaning of his team's accomplishments across to people in the wider computer world.

The augmentation center, as planned, had grown to seventeen people by 1968. They were on their third upgraded computer system, and the software was evolving from the first crude experimental versions to a real working toolkit for information specialists. In a matter of months, the SRI Augmentation Research Center was due to become the Network Information Center for ARPA's experiment in long-distance linking of computers—the fabled ARPAnet.

In the fall of 1968, when a major gathering of the computer clans known as the Fall Joint Computer Conference was scheduled in nearby San Francisco, Doug decided to stake the reputation of his long-sought augmentation laboratory in Menlo Park—literally, his life's work by that time—on a demonstration so daring and direct that finally, after all these years, computer scientists would understand and embrace that vital clue that had eluded them for so long.

Those who were in the audience at Civic Auditorium that afternoon remember how Doug's quiet voice managed to gently but irresistibly seize the attention of several thousand high-level hackers and engineers for nearly two hours, after which the audience did something rare in that particularly competitive and critical subculture—they gave Doug and his colleagues a standing ovation.

The audience, in the same room where the first "computer faire" for microcomputer homebrew hobbyists was held some years later, witnessed a kind of media presentation that nobody in the computer milieu had ever experienced before. State-of-the-art audiovisual equipment was gathered from around the world at the behest of a presentation team that included Stewart Brand, whose experience in mind-altering multimedia shows was derived from his production of get-togethers a few years before this, held not too far from this same auditorium, known as "Acid Tests."

Doug's control panel and screen were linked to the host computer and the rest of the team back at SRI via a temporary microwave antenna they had set up in the hills above Menlo Park. While Doug was up

Doug Engelbart at console on stage before the presentation ARC made in October, 1969, for the annual meeting of the American Society for Information Services. This presentation was similar to one made a year earlier at the Fall Joint Computer Conference. (Courtesy of Douglas C. Engelbart.)

there alone in the cockpit, a dozen people under the direction of Bill English worked frantically behind the scenes to keep their delicately transplanted system together just long enough for this crucial test flight. For once, fate was on their side. Like a perfect space launch, all the minor random accidents canceled each other. For two hours, seventeen years ago, Doug Engelbart finally got his chance to take his peers—augmentation pioneers and number crunchers as well—on a flight through information space.

Fortunately for the historical record, a film of the event was made. Those who were there at the original event say that the sixteen-millimeter film is a poor shadow of the original show. During the original presentation, an advanced electronic projection system provided a sharply focused image, twenty times life-sized, on a large screen. Doug was alone on the stage, the screen looming above and behind him as he sat in front of his CRT display, wearing the kind of earphone-microphone headsets that radar operators and jet pilots use, his hands resting on an unusual-looking control console connected to his chair.

The specially designed input console swiveled so he could pull it onto his lap. A standard typewriter keyboard was in the center, and two small platforms projected about six inches on either side. On the platform to his left was a five-key device he used for entering commands, and on the platform to the right was the famous "mouse" that is only now beginning to penetrate the personal computing market —a device the size of a pack of cigarettes, with buttons on top, attached to the console with a wire. Doug moved it around with his right hand.

In front of him was the display screen. The large screen behind him could alternate between, or share, multiple views of Doug's hands, his face, the information on his display screen, and images of his colleagues and their display screens at Menlo Park. The screen could be divided into a number of "windows," each of which could display either text or image. The changing information displayed on the large screen, activated by his fingertip commands on the five-key device and his motions of the mouse, began to animate under Doug's control. Everyone in the room had attended hundreds of slide presentations before this, but from the moment Doug first imparted movement to the views on the screen, it became evident this was like no audiovisual presentation anyone had attempted before.

Engelbart was the very image of a test pilot for a new kind of vehicle that doesn't fly over geographical territory but through what was heretofore an abstraction that computer scientists call "information space." He not only looked the part, but acted it: The Chuck Yeager of the computer cosmos, calmly putting the new system through its paces and reporting back to his astonished earthbound audience in a calm, quiet voice.

Imagine that you are in a new kind of vehicle with virtually unlimited range in both space and time. In this vehicle is a magic window that enables you to choose from a very large range of possible views and to rapidly filter a vast field of possibilities—from the microscopic to the galactic, from a certain word in a certain book in a certain library, to a summary of an entire field of knowledge.

The territory you see through the augmented window in your new vehicle is not the normal landscape of plains and trees and oceans, but an *informationscape* in which the features are words, numbers, graphs, images, concepts, paragraphs, arguments, relationships, formulas, diagrams, proofs, bodies of literature and schools of criticism. The effect is dizzying at first. In Doug's words, all of our old habits of organizing information are "blasted open" by exposure to a system modeled, not

The first "mouse," a breakthrough in computer input devices, constructed by Engelbart for an early ARC laboratory in mid-1960s. (Courtesy of Douglas C. Engelbart.)

on pencils and printing presses, but on the way the human mind processes information.

When the new vehicle for thought known as Arabic numbers was introduced to the West, and mathematicians found that they didn't have to fumble with Roman numerals in their calculations anymore, the mental freedom must have been dizzying at first. But not nearly as dizzying as this. There is a dynamism to the informationscape that needs no explanation, that needs only to be experienced to be understood. In that sense, Doug knew he had no choice but to take the risk of putting it up on the big screen and letting his audience judge for themselves.

Even the chewing-gum-and-baling-wire version Doug was attempting to get off the ground in 1968 had the ability to impose new *structures* on what you could see through its windows. The symbolic domain, from minutiae to the grandest features, could be rearranged at will by the informationaut, who watched through his window while he navigated his vehicle and the audience witnessed it all on the big screen. Informational features were reordered, juxtaposed, deleted, nested, linked, chained, subdivided, inserted, revised, referenced, expanded, summarized—all with fingertip commands. A document could be called up in its entirety, or the view could be restricted to only the first line or first word of each paragraph, or the first paragraph of each page.

One of the example tasks he demonstrated involved the creation of the presentation he was giving at that moment, from the outline of the talk to the logistics of moving their setup to Civic Auditorium. The content of the information displayed on the screen referred to the lecture he was giving at that moment, and the lecture referred to the information on the screen—an example of the kind of self-referential procedure that programmers call "recursion."

Doug moved his audience's attention through the outline by the way he manipulated their "view" of the information. His manipulations maneuvered the screen display and the audience's consciousness through categories of information, zoomed down into subcategories, broke them into their atomic components, rearranged them, then zoomed back up the hierarchy to meet the vocal narration at a key point in the story, when the words on the screen and the words coming from the narrator merged before branching off again. It was an appropriately dramatic presentation of a then-novel use of computers. While it appeared to be a radically sudden innovation to many of those in the audience, it was the culmination of careful experimentation at ARC that had already spanned most of a decade.

It is almost shocking to realize that in 1968 it was a novel experience to see someone use a computer to put words on a screen, and in this era of widespread word processing, it is hard to imagine today that very few people were able to see in Doug's demonstration the vanguard of an industry. When time-sharing systems first allowed programmers to interact directly with computers, in the early 1960s, the programmers developed tools known as "text editors" to help them write programming code. (The first one at MIT had a hand-lettered sign that dubbed it "expensive typewriter.") But "word processing" for nonprogrammers was still far in the future, despite Engelbart's demonstration of its potential.

The quality of video display technology in 1968 was also amazingly primitive by today's standards. The letters and numbers on Doug's screen looked as if they were handwritten—closer to the crude swaths "painted" onto a radar screen than the crisp pixels we are accustomed to seeing today on video display terminals.

In seeking a domain where a small success would mean a large boost in effectiveness, and where success would attract a large-scale research and development effort, Doug chose to augment the "humdrum but practical and important sorts of tasks" that occupy an increasing proportion of the people in our society: preparing, editing, and publishing documents. This area of document preparation and communication was but a small slice from the grand range of applications he envisioned, but it was one tool that the augmentation team itself needed immediately, and one that every laboratory and office in the world would want—as soon as people understood that computers weren't just calculators.

The seventeen members of the Augmentation Research Center, Engelbart explained during their 1968 show, were attempting to create a medium that would be useful to the other ARPA computer researchers and eventually to anyone who works with information. At the same time, this was a behavioral science experiment as well as a computer systems experiment, because the project team would be the subjects as well as the architects of the research. Making computers do what they wanted them to do was only the beginning. The really difficult work was adjusting themselves to new ways of working and thinking.

Consequently, one of the first projects was to create a system to make it easy for the members of the research team—and eventually for other intellectual workers—to compose, store and retrieve, edit, and communicate words, numbers, and graphics. "Text editing" had to become more amenable to nonprogrammers and more suited for the expression of thoughts and composition of prose.

They needed to invent the display devices and adapt the computer and write the programs; then they had to use what they had invented to compose a description of the system. The hardware and software specialists worked on representing symbols on screens and storing them in the computer's memory. Then the communications specialists used the text editors to write the manuals to instruct future members of the growing project in the use of the new tools.

The text-editing system was the first stage of Doug's long-term plan. The actual use of the system to design and describe the next generation was the second stage. Both stages were accomplished by 1968. Even as early as 1968, NLS was not limited to what we now call a word processing system. The third-stage goal was to build an entire toolkit for intellectual tasks, and develop the procedures and methods by which those tools could be used, individually and collectively, to boost the performance of people who did information-related work. The toolkit would then be used to develop new modes of computer-aided human collaboration.

Software was created to connect the text-editing system with a special kind of electronic filing arrangement that would serve as a unifying memory, record, and medium for their individual efforts. The software *journal* through which individuals and groups could have access to a shared thinking and communicating space had been in development since 1965–1966; it enabled individuals to insert comments into the group record of the augmentation experiments (or browse through them), and enabled programmers to trace the way system features had evolved. The journal, along with with *shared screen telephoning* to enhance real-time, one-to-one communications, was part of the overall *dialogue support system* designed to help increase effectiveness of group communication and decision making.

The idea of the journal predated the development of computer networks and teleconferencing, originating as it did with a dozen terminals connected to a single multiaccess computer. It was an important first try at "reaching through" the toolkit to engage in communication with another human user of the system. It was a theoretical precursor to the "electronic mail" medium that was to evolve when the ARPA network became operational in the early 1970s. When ARPAnet came

Doug Engelbart and project sponsors using augmentation technology to enhance a meeting in 1968. Engelbart is operating controls. In the foreground is Barry Wessler, then Bob Taylor's assistant in the ARPA Information Processing Techniques office. Behind Engelbart are Bill English, Don Andrews, and Dave Hopper. (Courtesy of Douglas C. Engelbart.)

along, connecting many computers in different locations into a shared computational "space," it wasn't such a shocking new medium to those few ARC pioneers who had been working on a smaller, localized version for years.

The journal was designed to bring order to the stream of dialogues, notes, and publications generated in the process of building the system and finding out how to work it. Besides serving as an electronic logbook that would be useful to human factors specialists and systems programmers, the journal was meant to be a medium for a formal dialogue among users that would serve the same purpose as today's traditional libraries and professional journals—but would do it in such an amplified manner that it would become a uniquely powerful method of transmitting knowledge.

For example, scientific journals in every field follow a form in which a paper describing research results is refereed, then published, after which subsequent papers can cite the previous paper. The record of any field of scientific knowledge—and the forum in which the significance of findings is debated—consists of a growing list of journal citations and accompanying text. It takes time for new information and comments to circulate, and it takes a relatively long time for individuals to thread their way through a history of branching citations. In the NLS version, it is very easy to jump directly and quickly from any article to the text of cited articles and back—reducing to seconds or minutes procedures that would take hours or months in even the most efficient traditional library/journal system.

Publication and distribution are radically changed by a computerized system, since it is so easy to automatically notify everybody on a certain kind of reading list that a new publication matching their interest profile is now available. Distribution lists can be members of distribution lists— you can designate a list to be the recipient of an announcement, and every member of the designated list will receive your message. Messages and articles can consist of lists of citations, and catalogs and indices can be message forms of their own. Ideas and hypotheses could be conveyed by telling interested members of the community to read a certain list of cited articles in a particular order.

This more formal and highly structured kind of intellectual discourse is essential to science, but is not the usual mode of communication used in the day-to-day affairs of ordinary citizens. As Licklider and Taylor, Doug's long-time colleagues and principal funders, pointed out in 1968, the new interactive computers and the new intercomputer networks would make it possible to use tools like NLS to construct a computer-aided *community* in which not only intellect but *communication* could be augmented.

At the most fundamental level, communication begins when two or more people need to share information, transact business, make decisions, resolve differences, reach agreements, solve problems, communicate plans. One of the early creations in the NLS collection of software levers and pulleys and skyhooks brought the other capabilities of the system to bear on communications. ARC developed a "mode of teleconferencing" whereby: [8]

> . . . two or more people, positioned at separated display consoles, can link their displays so that all see the same image, and at option any can exercise control. When simultaneously talking on the telephone the resulting dialogue can be uniquely effective—corresponding to an in-person conference around a collective assemblage of their scratch pads, working records, and individual support facilities. . . .
>
> But consider the great potential already existing when some of the participants—or even a single participant—can effectively use computer tools to work with the relevant materials and processes. There is great value in merely conducting themselves as though they were congregated at a magic blackboard—each easily able to pull forth materials from his notes or familiar reference sources, copy across into his private workplace any material offered from what the other brings forth.

In 1969, ARC became one of the original nodes of the ARPAnet system that connected defense-related research computers around the country into a network. The network, Bob Taylor's brainchild, used common-carrier communication lines to interconnect computers in different parts of the country. While the separate time-sharing communities were busy exchanging data, programs, and messages, the ARC people saw their participation in the network as an opportunity to put their knowledge to good use, and to extend their experiment beyond their SRI laboratory to include everyone around the country who was connected to the network.

As the network grew, ARC branched out from its primary activity of continually redesigning itself. It began serving as the Network Information Center, offering referencing and organizing services for the distributed community of ARPAnet users. No longer languishing in a half-forgotten Quonset somewhere on the huge SRI grounds, the augmentation laboratory, equipped with the latest time-sharing hardware, was by 1970 the proud subject of VIP tours.

After so many years of solitary envisioning, Engelbart had become even more optimistic about the ultimate significance of their enterprise than he had been when he started. In the spring of 1970 he told his colleagues at the Interdisciplinary Conference on Multi-Access Computer Networks: [9]

> . . . It has been my business to struggle with these concepts for two decades now, and the signs that I read at least tell me that the changes in our ways of thinking and working will be more pervasive and extreme than ANY OF US appreciates—a revolution like the development of writing and the printing press lumped together. . . .
>
> It will take the explorers of this domain decades to even map its currently visible dimensions. The real rush hasn't begun: this Conference is a meeting of suppliers looking at the prospector trade; we haven't really been giving attention to the developments that will follow the prospecting.
>
> My research group is now moving into a next stage of work that we call "team augmentation." Here, instead of just the individual facilitating his private domain of searching, studying, thinking and formulating, as his office place provides for him, we are exploring what can be done for a team of "augmented individuals" who have in common a number of terminals, a set of computer tools, working files, etc. (as we do), to facilitate their team collaborations.

The problem-solving assistance Engelbart had dreamed about alone in the 1950s became the "integrated working environment" he proposed in 1963, which in turn grew into the toolbuilders' toolkit that he and his small group of colleagues used to build an "intellectual workshop" throughout the remaining seven years of the decade. By the early 1970s, the wider community of ARPA-funded computer researchers and representatives of the business world were joining the bootstrapping process. Paradoxically, just when their leader decided that "team augmentation" would be their goal, his own team began to react negatively to growing pressures—technological, psychological, and social.

Doug had always warned that "the larger augmentation system is much more complex than the technological 'subsystem' upon which it depends," and the 1970s were the era when ARC began to practice what Engelbart had preached. During the first decade of the laboratory's existence, computer technology had progressed at an astonishing pace, and the SRI crew were doing their utmost to use the innovations as quickly they came along.

The "rule of two" (that computer power would double every two years) and the Engelbart-induced zeal of the augmentation team kept them fueled for an effort to bootstrap and continually adjust themselves to the capabilities of their upgraded tools—an effort that required extraordinary intensity. The bootstrapping and readjusting continued with unabated enthusiasm, at least until the early 1970s, when the idea of building a system that was meant to "transcend itself every six to eight months" to keep pace with hardware and software advances turned out to be more pleasant to contemplate than to carry out. It had been challenging and exhilarating to build this new system for augmenting thought— but it wasn't as much fun having one's work habits augmented at a forced-march pace.

When both the old-timers and newcomers to the growing project faced the task of learning new roles, changing old attitudes, adopting different methods, on a regular basis, just because the system enabled them to do so, the great adventure became more arduous than any of the ARC pioneers/experimental subjects had anticipated. So a psycholo-

Two workstations in ARC's bullpen, 1973. In the foreground, seated, is Smokey Wallace, now short-haired, with neatly trimmed beard, a researcher and executive with Adobe Systems. Leaning over, pointing to the screen, is Bill Ferguson, now an attorney. (Courtesy of Douglas C. Engelbart.)

gist was brought in to consult about those parts of the system that weren't to be found in circuitry or software, but in the thoughts and relationships of the people who were building and using the system.

Dr. James Fadiman joined ARC as an observer-catalyst-therapist. Fadiman was particularly interested in the ways human consciousness and behavior change in new situations, and it didn't take him long to realize that the process of "being augmented" was in fact a new, nonchemical form of altered consciousness.

Several of the things Fadiman learned about the "augmentation experience" have taken more than a decade to filter out to the people who design computers for nonexperts. One thing he learned almost immediately was that most people resist change, especially in the workplace, and the resistance works both ways—people who are resistant to learning an augmentation system are equally resistant to giving it up once they have adopted it. The initial resistance is partially grounded in a general fear of the unknown.

Doug Engelbart, of course, saw things on his own scale, and through the eyes of an engineer. There would be rough spots, software and interpersonal bugs, arguments and conflicts, to be sure—but the master plan was progressing nicely, considering all those years he had worked alone. The toolkit had become a workshop, and they knew that the workshop indeed worked because they had been their own guinea pigs for a decade.

In the same 1970 address in which he referred to the multiaccess computing community as a "meeting of suppliers looking at the prospector trade," Engelbart also predicted that the future would see "a steadily increasing number of people who spend a significant amount of their professional time at terminals," and speculated that the future of dispersed personal augmentation systems linked together into network communities would create new kinds of societal institutions: "In particular, there will emerge a new 'marketplace,' representing fantastic wealth in commodities of knowledge, service, information, processing, storage, etc."

In his usual forge-ahead manner, Engelbart was already bringing members of the business community into the ARC experiment. Business managers and management scientists had been working at ARC, experimenting with using NLS tools to manage the steadily growing ARC project. In proper bootstrapping style, they looked at their attempts to apply the system to their own research management as yet another experiment. Richard Watson and James C. Norton worked closely with ARC to develop their experimental discoveries into a system that would be usable by people who were not computer experts but whose occupations involved the manipulation of information.

Sometime in the early 1970s, Engelbart was inspired by a book, just as he had been enthused by magazine articles by Bush and Licklider in years past. This time, it was the theory proposed by business management expert Peter Drucker in the late 1960s. Knowledge, by Drucker's definition, is the systematic organization of information; a knowledge worker is a person who creates and applies knowledge to productive ends. The rapid emergence of an economy based primarily on knowledge, Drucker predicted, would be the most significant social transformation of the last quarter of the twentieth century.[10,11]

Drucker noted something about the future of knowledge in the American economy that seemed to converge, from an unexpected but not unpredictable direction, with the course Engelbart had plotted for the augmentation project at the beginning of its second decade. Drucker was one of the first of a growing number of social scientists who have claimed that an examination of labor statistics reveals a great deal about the role of knowledge work in everybody's future.

In 1973, ten years after his solo "Framework . . . ," Engelbart, Watson, and Norton presented a paper on "The Augmented Knowledge Workshop" to the National Computer Conference. Acknowledging their debt to Drucker's ideas, the authors pointed out that the special computer systems that had been evolving at ARC were designed to alleviate the problems associated with "the accelerating rate at which knowledge and knowledge work are coming to dominate the working activity of our society": [12]

> In 1900 the majority and largest single group of Americans obtained their livelihood from the farm. By 1940 the largest single group was industrial workers, especially semiskilled machine operators. By 1960, the largest single group was professional, managerial, and technical— that is, knowledge workers. By 1975–80 this group will embrace the majority of Americans. The productivity of knowledge has already become the key to national productivity, competitive strength, and economic achievement, according to Drucker. It is knowledge, not land, raw materials, or capital, that has become the central factor in production.

Noting Drucker's use of terms such as "knowledge organizations" and "knowledge technologies," Engelbart, Watson, and Norton specified an augmented knowledge workshop that was nothing less than a totally redesigned working environment for everybody in the "knowledge sector." The authors acknowledged that ordinary knowledge workshops— offices, boardrooms, libraries, universities, studios—have existed for centuries. Augmented knowledge workshops, however, existed only as proto-types, and would not come into widespread usage until the technologies pioneered at ARC (and by then, at a new place across the creek, called PARC) grew economical enough to sell as office equipment. This was the origin of an idea that was later adapted by others in a truncated version known as "The Office of the Future."

The authors described the technology they had built and used for augmenting their own knowledge work as individuals and in groups, but emphasized that the tools were only the first part of a total transforma-tion of the system—including changes in methods, attitudes, roles, life-styles, and working habits. They knew from their own experience that the psychological and social adjustments would be the most intense and volatile changes set off by the introduction of these systems into existing organizations.

In 1975, after twelve years of continuous funding, ARPA dropped ARC. The staff quickly shrank from a high of thirty-five to a dozen, then down to a few, and finally down to Doug Engelbart and a large amount of software. A decade of useful work is an unheard of length of time in the hyperaccelerated world of software technology, but boot-strapping had kept NLS continually evolving as it expanded its usefulness, as it was moved up to machines with larger memories and faster pro-cessors, and as the community of users thought of new things to do with it.

Even before ARPA drastically reduced its funding, ARC had started a subscription service to several corporations who wanted to experiment with using the services of the augmentation system. The way Engelbart saw it, it was time to bring the system out of the research world, after its extended gestation, to test it on a community of real-world users. The way SRI probably saw it was that the whole project was obviously finished as a magnet for research funds, and they might as well sell it, since they never did understand it. In 1977, SRI sold the entire augmenta-tion system to Tymshare Corporation, and Engelbart went with it. The system, renamed "Augment," is now marketed by Tymshare as one of their office automation services.

Nobody disputes that Engelbart's vision was the single factor that stayed stable through twenty of the most turbulent years of computer science, and those few colleagues who know of his importance to the evolution of computing are loathe to speak unkindly of him, yet the tacit consensus is that Doug Engelbart the visionary allowed himself to remain fascinated by an obsolescent vision. NLS was powerful but very complex, and the notion of a kind of knowledge elite who learned complex and difficult languages to operate information vehicles is not as fashionable in the world of less sophisticated but more egalitarian personal computers created by Engelbart's students.

The twelve years of ARC's heyday at SRI, from 1963 to 1975, were technologically wild years. That period was one of enormous historical, social, and cultural upheavals, as well. Mistakes, conflicts, blind alleys, and other pitfalls were unavoidable during the course of a project that began in the Kennedy administration and continued throughout the years of the Vietnam war, campus revolts, assassinations, the emergence of the counterculture, the advent of women's liberation, Watergate, and ended during the Carter administration.

As individuals, and as a group, ARC wasn't immune to the conflicts that affected the rest of the culture, although it was privy to its own mutated forms of it. Before the counterculture made its media splash and thousands of affluent American offspring started acting weird and growing their hair long, places where powerful computers were to be found had already spawned their own brand of weirdo—the hacker. The advent of this new subculture within the computer subculture was not the direct cause of ARC's downfall, but it was symptomatic of the problems Engelbart faced in the 1970s.

Engelbart found himself caught between the conservatism of his employers and the radicalism of his best students. ARC had always seemed a bit strange to the old-line data-processing types at SRI, and these new people hanging out at Doug's lab added cultural as well as technological differences to an already strained relationship. To say that SRI is conservative is an understatement. Although some of the subjects their researchers pursue can be unorthodox, their clients are such straitlaced institutions as the Defense Department, the intelligence community, and the top one hundred corporations.

Hackers were barely tolerated in the long, clean, high-security halls of SRI. But when the counterculture started to infiltrate, and the rumors started about some of the hackers augmenting their consciousness in more ways than one, SRI brass became extremely uncomfortable.

There was trouble from within, as well as from above. Some of the experiments in "new-age" social organization, encouraged by Engelbart himself, threatened to split the ARC group into two camps—those who were still techies at heart, concerned only with the advancement of the state of the computing art, and those who saw augmentation as an integral part of the wider countercultural revolution that was going on around them. And there were those who felt that even Doug's technological ideas, although they might have once been radical and futuristic, were becoming outmoded. The idea of augmentation teams and high-level time-shared systems began to seem a bit old-hat to the younger folks who were exploring the possibility of personal computers.

In the early 1970s, some of Engelbart's first and most important re-cruits, who had helped him create the first NLS system, left SRI for PARC, the new research center Xerox was putting together. The new Xerox facility was a hotbed of augmentation-oriented thought, but with a major difference—the advent of large-scale integrated circuitry made it possible to dream of, and even design, high-powered computers that could fit on an individual's desk. This emphasis on one person, one computer made for important philosophical and technical differences with Engelbart's approach.

For a while, Engelbart at SRI and his former students at Xerox were engaged in collaboration, but eventually PARC and ARC drifted apart. Doug still dreamed of creating augmentation centers in universities and industries, providing a service for any team of people who worked with information. The former ARC members were looking forward to an even wider potential computer-using population. The idea at Xerox was to use the new integrated circuit technology to create computers more powerful than the previous generations of minicomputers—and to devote an entire computer to each person, instead of sharing it among thirty or forty users.

PARC, as we shall see, went on to become the new mecca for those who saw the computer as a tool for augmenting human intellect. ARC never seemed to make it to the promised land, and the former point-man for radical technology seemed to be more and more isolated in an interesting but less than influential backwater. As more and more of Engelbart's earlier dreams became realities in other institutions, this judgment seemed to be less than fair. It is impossible to tell if there would have been a PARC if there hadn't been an ARC, and while the miniaturization revolution made personal computers inevitable in a technical sense, there is good reason to question whether the kind of personal computing that exists today would ever have been developed if it had not been for the pathfinding work accomplished by Engelbart and his colleagues.

Doug Engelbart and the people who helped him build ARC did not succeed in building a knowledge workers' utopia. Some hackers do seem to be pathologically attached to computers. These facts might have very little to do with the way other people will use descendants of the tools they created. In fact, if you think about it, some of the wildest and woolliest of the MAC and ARC hackers were following in a long tradition of people who weren't exactly run-of-the-mill citizens—from Babbage and Lovelace and Boole to Turing and von Neumann.

It must be remembered that MAC and ARC were only part of a larger effort to raise computing to a whole new level, and hackers weren't the only scientist-artisans involved in that effort. Whatever future historians decide about the personalities of the people involved in carrying out this unprecedented exercise in planned breakthrough, they will have to consider the role of the hackers who created time-sharing, computer networks, and personal computers in the 1960s and early 1970s, not out of sick obsession or in-group frivolity, but out of a serious desire to construct a new medium for human communication.

For the time being, Doug Engelbart still works away at his original goals, adapting the core of NLS to the new kind of computers that have come into use in the 1980s. To Tymshare Corporation's customers, the Augment system seems less science-fiction-like and more practical in this age of office automation. People in the business world are beginning to pay attention to what Doug is saying, for the first time since he started saying it, decades ago.

Still, Doug is neither rich nor famous nor powerful—not that these were ever his goals. All he seems to hunger for is all he ever hungered for—a world that is prepared for the kind of help he wants to give. Ironically, his office at Tymshare in Cupertino, California, is merely blocks away from the headquarters of Apple Corporation, where icons and mice and windows and bit-mapped screens and other Engelbart-originated ideas are now part of a billion-dollar enterprise.

CHAPTER TEN

The New Old Boys from the ARPAnet

Bob Taylor's office window at Xerox Corporation's Palo Alto Research Center (PARC) overlooked the red-tiled towers of Stanford and the flat roofs of research parks stretched out to the horizon. The electronic window next to his desk overlooked another kind of world. While he started talking to me, he was also interacting with colleagues in his building and elsewhere in the global information community.

In 1983, it was not unusual to see an executive, especially a manager in a computer research organization, using a personal computer in his office. The unique thing about *this* personal computer was that it was an *Alto*—the *first* personal computer. Taylor and his group had been using it since 1974. A small cable connected the Alto to the *Ethernet*— a medium that linked the researchers at PARC with each other and with colleagues around the world.

The screen was taller than most computer displays, and it looked different from other computer screens, even when seen from across the room. Instead of a single screen-sized frame filled with numbers or letters or graphs, there were a number of squares of various sizes, known in Xerox parlance as *windows*, that looked like overlapping pieces of paper on a desk. The symbols and images were also distinctly sharper than what I was accustomed to seeing on a computer screen.

Xerox PARC's Alto workstation. With its bit-mapped, high-resolution screen, mouse pointing device, large internal and external memory storage, and special software, the Alto was the first true personal computer. (Courtesy of Xerox Corporation.)

The mouse, an update of Engelbart's innovation, was connected to the Alto with a thin wire. As Taylor slid the mouse around the desk surface next to the screen, a small dark pointer shaped like an arrow moved around the screen. When he clicked one of the buttons on top of the mouse or moved the pointer into a margin, the pointer changed shape and things happened on the screen. In 1984, Apple Corporation's Macintosh computer introduced a mass market to this way of handling an electronic desktop. To Taylor, it wasn't particularly futuristic. Altos and Ethernets had been in operation since 1974 around here.

By 1983, Bob Taylor was only half-satisfied with his progress toward what he and a few others set out to achieve twenty years ago, because he believed that the new technology was only halfway built. Despite the fact that the office he was sitting in, the electronic workstation at his fingertips, and the research organization around him were functioning examples of what the augmentation community dreamed about decades ago, Taylor thought that it might take another ten or twenty years of hard work before the interactive informational communities foretold by Bush and Licklider would truly affect the wider population.

In 1965, at the age of thirty-three, Robert Taylor worked out of his office in the Pentagon, as deputy director, then as director, of the ARPA Information Processing Techniques Office. His job was to find and fund research projects involving time-sharing, artificial intelligence, programming languages, graphic displays, operating systems, and other crucial areas of computer science. "Our rule of thumb," he remembers, "was to fund people who had a good chance of advancing the state of information processing technology by an order of magnitude."

Bob Taylor was also responsible for initiating the creation of the ARPAnet—the prototype network community of computers (and minds) created by the Department of Defense, an effort that began in 1966 and became an informal rite of passage for the nucleus of people who are still advancing the state of the computing art. Larry Roberts, who was responsible for getting the network up and running, succeeded Taylor when Taylor left ARPA in 1969. After a year at the University of Utah, Taylor joined the research effort Xerox Corporation was assembling near Stanford.

In 1970, a combination of growing opposition to the Vietnam war, and the militarization of all ARPA research, meant that an extraordinary collection of talent in the new fields of computer networks and interactive computing were looking for greener pastures at a time when one corporation decided to provide the greenest pastures imaginable.

In 1969, Peter McColough, CEO of Xerox Corporation, announced his intention to make Xerox "the architect of information" for the future. To that end, a research organization was assembled in Palo Alto, in the early 1970s. McColough put a man named George Pake in charge. One of the first things Pake did was hire the best long-term computer visionary, research organizer, and people-collector he could find—Bob Taylor. At first, the newly recruited engineers, hackers, and visionaries worked in temporary quarters located in the Palo Alto flatlands, near the Stanford University campus. In the mid-1970s construction began on a prime piece of high ground above Hewlett-Packard, next to Syntex, in that fertile futurist enclave known as "The Stanford Industrial Park."

If ever there was a model environment for the technological cutting edge of the "knowledge sector," PARC was it. From the physicists in the laser laboratories and the engineer-artisans in the custom microchip shops to the computer language designers, artificial intelligence programmers, cognitive psychologists, video jockeys, sound engineers, machinists, librarians, secretaries, cooks, janitors and security guards, you got a nice, relaxed, model-utopian feeling from everybody you encountered.

The physical plant itself is an inescapable exercise in innovation. It took me a while to stop thinking of the place as being upside down. Since the terraced glass-and-concrete structure was built halfway embedded in Coyote Hill, Zuni Pueblo style, the main entrance is on the top floor. To get to the second floor from the ground floor, you go *down*. The linked quadrangles of offices, laboratories, and meeting rooms

Xerox Corporation's Palo Alto Research Center, known as PARC, built into Coyote Hill in the Stanford Industrial Park. In the 1970s, an extraordinary nucleus of people gathered at PARC and created the foundations of personal computer technology. (Courtesy of Xerox Corporation.)

wind around atriums and gardens. The cafeteria overlooks Palo Alto; you can take your tray out to the terrace and look down on the bay from the vantage of this twenty-first-century cliff dwelling.

Off the corridors that wind around the quadrangles are office cubicles, many with their doors open. Inside the open cubicles, various people talk on telephones or stare at their distinctively oblong Alto screens. Some cubicles have plants, posters, bean-bag chairs, stereos, bicycles. They all have bookshelves with rows of books and the bright blue and white binders used on the reports PARC publishes for the outside world. Many of the cubicle dwellers are young. A larger proportion of them than you might suspect are women. It has always been a multinational-looking crowd.

I had no problem distinguishing Taylor from the assorted scientists, engineers, professors, hackers, longhairs, and boy and girl geniuses around him. The few differences in style were subtle but visible, nevertheless. While many of his colleagues opt for sandals, down jackets, techno-hippie ponytails, blue jeans, and rumpled cords with or without bicycle clips, Taylor is likely to be found in a pressed tweed jacket and unrumpled slacks. His blond hair is casual but neat. When he's trying to see if you are following his line of thought, he tilts his forehead in your direction and targets you with pale blue eyes over what would pass for granny glasses if his shirt were denim instead of oxford cotton. He smiles often, sometimes as a form of punctuation. A trace of Texas drifts into his voice at times.

It is Taylor's belief that the idea of personal computing was a direct outgrowth of what Licklider started in the early 1960s with time-sharing research. Time-sharing, like the first high-level languages, was a watershed for computer science and for the augmentation approach. It also created a new subcommunity within the computation world, a community of interests that cut across the boundaries of military, scientific, academic, and business computing. It was a relatively small subculture within the larger community of computer scientists and computer systems builders. They were bonded by a common desire for a certain kind of computer they wanted for their own use, and by a decade of common experiences as part of the ARPA research effort to build the kind of computers they were then using. Many of the time-sharing veterans who started out as undergraduate hackers at Project MAC or as ARPA-funded engineers in Berkeley and Santa Monica were to meet later, in the research sanctums of Bell, SRI, Rand, and (mostly) at PARC.

Time-sharing was an early and effective application of the philosophy that the existing means of using computers should be tailored to the way people function, rather than forcing the people who want to use them to conform to mechanical constraints. Without the development of multiaccess computing in the early sixties, the idea of personal computing would never have become more than a dream.

In the early 1960s, data processing was what one was expected to do with a computer, and one hardly ever did it directly. First, a program and its raw data had to be converted into a shoebox full of punched cards. The cards were delivered to a data processing center, where a systems administrator decided how and when they were to be fed to the main computer. (These fellows were, and still are, a rich source of anecdotes in support of the "programming priesthood" mythology.) You came back an hour or a day or a week later and retrieved a thick printout and a hefty bill. The keypunch-submit-wait-retrieve ritual was called "batch processing."

By 1966, groups in California and Massachusetts were well on the way toward raising the art of computer programming to a high enough level to do some truly interesting things with computers. Licklider and a few others suspected that if they could make the power of computers more directly accessible to the people writing and running programs, programmers might be able to construct new and better kinds of software at far greater speed than heretofore possible.

Among the capabilities that came with the increasingly sophisticated electronic hardware and software were powers to model, represent, and search through large collections of information. With sufficient speed and memory capacity, computers were gaining the power to assist the creative aspects of communication. But serious obstacles had to be overcome to bring that power out where people could use it.

It is hardly possible to interact dynamically with your program when you have to dump boxes of punchcards into card-readers, then decipher boxes of printout. Since a large part of the process of building a program is a matter of tracking down subtle errors in complex lists of instructions, the batch processing ritual put an effective limit on how much programmers could do, how fast they could do it, and the quality of the programs they could produce.

Batch processing created two problems: The computers could handle only one program (and one programmer) at a time, and programmers weren't able to interact directly with the computer while their programs were running. Time-sharing was made possible because of the enormous gap between the speed of computer operations and the rate of information transfer needed to communicate with a human. Even the fastest typist, for example, can enter only a single keystroke in the length of time it would take the computer to perform millions of operations. Time-sharing gives each of the 20, 50, or 100 or more people who are using the computer the illusion that he or she has the computer's exclusive "attention" at all times, when in reality the computer is switching from one user's task to another's every few millionths of a second.

When the first programmers gained interactive access to the computer, they also gained a new freedom to create ever more powerful programs and see the results much more quickly than ever before. Programmers of the first multiaccess computers of the sixties were able to submit programs a piece at a time and receive responses a piece at a time, instead of trying to make the whole programming job work, for better or worse, in a single batch. By eliminating the "wait and see" aspect of batch-processing, time-sharing made it possible for programmers to treat their craft as a performing art.

"When I became director of the ARPA Information Processing Techniques Office, the time-sharing programs were already running," Taylor recalls, "but they weren't complete, so the work continued on time-sharing while I was director. It was clear, though, that this was an important breakthrough in information processing technology, so I became involved in the technology transfer between the different experimental systems, and eventually to military and civilian computer applications.

"We came up against some rigid attitudes when we talked to many people in the industry. IBM ignored the ARPA stuff at first. They simply didn't take it seriously. Then GE agreed to cooperate with MIT and Bell Laboratories to develop and market a large time-sharing system. IBM said, 'Whoops, something's happening here,' and they went off with a crash project to retrofit one of their 360 systems to time-sharing. They took orders for a few and the system bombed. They couldn't make the software work because they hadn't been down some of the same roads that the ARPA funded groups had been down years before."

Time-sharing research caused a kind of schism in the corporate research field. The first-generation priesthood seemed to be missing out on the inside action, for a change. Companies that paid attention to the time-sharing experience gained in the long run. It made Digital Equipment Corporation the "second name" in the industry. DEC paid attention to the ARPA-funded work and hired people when they got out of school, and profited from time-sharing.

The first thing Taylor went after, once the time-sharing project was on its way to completion, was a way of interconnecting the time-sharing communities. He had a privileged overview of the then-fragmented computer research world, since a good deal of his time was spent traveling to universities and think tanks, finding and funding researchers. Progress in the separate subfields of computer research was accelerating through the early 1960s. By 1966, the time was approaching when the pieces of the puzzle would be ready for assembly, and the separated teams would have to be in closer communication.

"Within each one of the time-sharing communities people were doing a variety of different kinds of computer research," says Taylor, "so the overall project of making the time-sharing system itself work was much more global than any one of the individual research fields that were being explored by different members of the time-sharing community— AI research, computer hardware architecture, programming languages, graphics, and so forth.

"We were surprised time and again by applications of the time-sharing system that nobody planned but someone invented anyway. The ability to share files and resources within a time-sharing system was one difficult problem to be solved. On the way to solving it, people discovered a new way of communicating with each other—something that was unexpected and became a unique medium of expression in the research community." Fifteen years since computer jockeys started having fun with it, that medium has become the commercial version known as "electronic mail."

Taylor saw the necessity of connecting to one another those isolated research communities that Licklider had seeded and Sutherland had nurtured. Many of the people in related fields but different institutions knew of each other, and many more did not. By 1965–1966, ARPA was supporting most of the nonindustrial computer systems research in the country, and thus Bob Taylor and his colleagues had a more up-to-date and comprehensive picture of the state of computer research than any of the individual researchers.

The people Taylor funded then undertook the planning and creation of a *network* of computers, located in different parts of the country, linked by common-carrier communication lines, capable of sharing resources and interacting remotely with the growing community of computer researchers. The people who were to build and ultimately make use of the system began to get together in person to talk about the technology needed to link resources in the manner they envisioned. Instead of working in isolation, a small group of leaders from the time-sharing research effort began to work in concert to design the first online, interactive communities.

A truly interoperating community capable of freely sharing resources across the boundaries of individual machines or geographical locations was more difficult to bring into existence than is suggested by the simplified general idea of plugging computers together via telephone lines. Very serious hardware and software problems had to be solved, and the "user interface" where the person meets the machine had to be further humanized.

Every year, starting in 1966, following the tradition established by Licklider and Sutherland, Taylor called a meeting of all the principal investigators of all his projects. It would be held in a dramatic place far removed from the usual locales of Cambridge, Berkeley, or Palo Alto. With these meetings, Taylor, who was neither an engineer nor a programmer (he was, in fact, a philosophy major and an experimental psychologist by training), began the all-important mixing and sifting of ideas he knew would be necessary to the cohesion of such a large, dispersed, and ambitious project.

"I constructed the meetings so they all had to get to know one another and argue with one another technically in my presence," Taylor recalls. "I would ask questions that would force people to take sides on technical issues. Lasting friendships were built from that give and take. I asked them difficult questions. They became comfortable asking one another difficult questions. Then, after they went back to their laboratories and campuses, their communications increased in both quality and quantity, because they knew each other."

Taylor also initiated annual conferences of graduate students. The best graduate students of the ARPA researchers had meetings of their own, away from the "older" folks like Taylor, who was, after all, in his midthirties. Like the bands of roving builders who planned the Gothic cathedrals of Europe, many of the computer-system builders who participated in the ARPA grad students' meetings were to meet again later at SAIL (Stanford Artificial Intelligence Laboratory) and PARC, and later still at Apple and Microsoft.

Taylor's idea of connecting the researchers by connecting their computers was inspired by a phrase he read in one of Licklider's 1966 papers, in which he proposed the idea of a very large-scale time-sharing system that he called "an intergalactic network." Taylor took it a step farther: If you could build a communication network, why not a computer network?

Instead of building larger numbers of longer-range communication lines between terminals and their central time-sharing systems, Taylor saw potentially greater benefits in creating a technology for different time-sharing systems to communicate with each other over long distances. Taylor sold ARPA on the idea, then hired a young Lincoln Lab researcher named Larry Roberts as project manager. The meetings and separate research projects continued for three years, before the first bits were sent over the ARPAnet in 1969. By this time, Taylor's opposition to the Vietnam war was growing, and he was reasonably certain that the project he had initiated was coming to completion, so he left ARPA.

While the number crunchers, batch processors, and electronic book-keepers continued to hold sway over the computer industry, the core members of the interactive computing community were beginning to experiment with their computers—and with themselves—through this unique new prototype of an interconnected computer community. It quickly turned out, to the delight of all the participants and to nobody's surprise, that the experimental network was evolving into a stimulating environment for communicating and sharing research information and even for transporting and borrowing computer programs.

The implications for human communication that were beginning to emerge from the experience of this computer-connected research community were discussed in an article published in April, 1968, titled "The Computer as a Communication Device." The principal authors were none other than J.C.R. Licklider and R. Taylor.[1]

Although the Department of Defense had an obvious interest in foster-ing the development of the technology they created in the first place, and the interconnection of computers had certainly become a necessity in conducting advanced weapons research, Licklider and Taylor were not talking about applying the network idea to the Strategic Air Com-mand or nuclear weapons research, but to the everyday communications of civilians.

The authors emphasized that the melding of communication and com-putation technologies could raise the nature of human communication to a new level. They proposed that the ability to share information among the members of a community and the presence of significant computational power in the hands of individuals were equal components of a new communicating and thinking environment they envisioned for the intermediate future. The implications were profound, they felt, and not entirely foreseeable: "When minds interact, new ideas emerge," they wrote.

The authors did not begin the article by talking about the capabilities of computers; instead, they examined the human function they wished to amplify, specifically the function of group decision-making and prob-lem-solving. They urged that the tool to accomplish such amplification should be built according to the special requirements of that human function. In order to use computers as communication amplifiers for groups of people, a new communication medium was needed: [2] "Creative, interactive communication requires a plastic or moldable medium that can be modeled, a dynamic medium in which premises will flow into consequences, and above all a common medium that can be contributed to and experimented with by all."

The need for a plastic, dynamic medium, and the requirement that it be accessible to all, grew out of the authors' belief that the construction and comparison of informational *models* are central to human communication. "By far the most numerous, most sophisticated, and most important models," in Licklider's and Taylor's opinion, "are those that reside in men's minds."

Collections of facts, memories, perceptions, images, associations, predictions, and prejudices are the ingredients in our mental models, and in that sense, mental models are as individual as the people who formulate them. This essential privacy and variability of the models we construct in our heads create the need to make external versions that can be perceived and agreed upon by others. Because society, a collective entity, distrusts the modeling done by only one mind, it insists that people agree about models before the models can be accepted as fact.

The process of communication, therefore, is a process of externalizing mental models. Spoken language, the written word, numbers, and the medium of printing were all significant advances in the human ability to externalize and agree upon models. Each of those developments, in their turn, transformed human culture and increased our collective control over our environment. In this century, the telephone system added a potent new modeling medium to the human communication toolkit. Licklider and Taylor declared that the combination of computer and communication technologies, if it could be made accessible to individuals, had the potential to become the most powerful modeling tool ever invented.

As an example of how a prototype computer communication system could be used to boost the process of group decision-making, Licklider and Taylor described an actual meeting that had taken place on just such a system. It was a project meeting involving the members of a computer-science research team. Although all the participants in the meeting were in one room, they spent their time looking at their display screens while they talked. A variety of diagrams, blocks of text, numbers, and graphs passed before their eyes via those screens.

The facility was, in fact, Doug Engelbart's Augmentation Research Center. The machine in another room that made the meeting possible was the latest kind of multiaccess computer that the time-sharing research of the last few years had produced.

Using the project meeting as a model, Licklider and Taylor showed how computers could handle the informational housekeeping activities involved with a group process. More importantly, they demonstrated how this subtle kind of communication augmentation could enhance the creative informational activity that took place. The ability to switch from microscopic details to astronomical perspectives, to assemble and reassemble models, to find and replace files, to cut and paste and shuffle, to view some information publicly and make private notes at the same time, to thumb through the speaker's files or check his references while he is talking, made it possible for people to communicate with each other through this computer system in a way not possible in a nonaugmented meeting.

"In a few years," the authors predicted, in the very first words of their article, "men will be able to communicate more effectively through a machine than face to face." Referring to their model technical meeting at SRI, Licklider and Taylor estimated that "In two days, the group accomplished with the aid of a computer what normally might have taken a week." [3]

This small group—the people together with the hardware and software of a multiaccess computer—constituted what Licklider and Taylor identified as one node of a larger, geographically distributed computer network. The key idea, Taylor and Licklider now recall, had been proposed by Wesley Clark in a cab ride to Dulles Airport, after a 1966 meeting about the network Taylor was trying to put together. The problem lay in deciding which levels of the existing computer and communication systems had to be changed to couple incompatible machines and software.

Many of the planners believed that a huge "host" computer in the center of the country would have to be specially designed and programmed to act as a switchboard and translator. Clark suggested that a small, general-purpose computer at each node could be turned into a "message processor." Through long-distance common-carrier communications, these "interface message processors" (known eventually as "imps") and their local multiaccess computer communities could be integrated into a kind of supercommunity.

The imps would take care of all the behind-the-scenes traffic controlling and error-checking functions needed to ensure accurate transmission of data—a significant task in itself—so the individual users wouldn't have to worry about whether the files they want to read or the programs they need to use are a thousand miles away or down the hall.

The resulting communication system became part of a new kind of computing system that was not confined to any single computer. Teams of ARPA-supported scientists found that they could invoke the use of a program residing in a computer located in Berkeley, California, feed the program with data stored in Los Angeles, then display the result in Cambridge, Massachusetts. The network was suddenly more important than the individual computers, as the computers became "nodes" in a geographically distributed supercomputer.

It began to be possible to think of a computer network that was not centrally controlled from any one place, in which the traffic control and data communication and behind-the-scenes number crunching required were invested in the software instead of the hardware. Instead of a huge host computer in the center of it all that received a stream of information from one computer, translated the stream into a form that could be decoded by another computer, and relayed the translated information to the receiving computer, the smaller imps at each node would accept and pass along information packets that had been translated into a common format by the imp connected to the originating computer.

The controlling agent in a "packet-switched" network like the ARPA-net was not a central computer somewhere, nor even the "message processors" that mediated between computers, but the packets of information, the messages themselves. Like the addresses on letters, the code by which information was packaged for transmission put into each packet all the information necessary for getting the message from origin to destination, and for translating between different kinds of computers and computer languages.

While the networking technology was evolving rapidly the number of computer terminals proliferated and the accepted way of using computers was beginning to change. By 1968, the punchcards and printouts of 1960 were being replaced by ever-more interactive means of communicating with the computer: a keyboard and teletype printer and, in some exotic quarters, a graphic display screen were becoming standard input and output devices for programmers.

To old-liners who were used to submitting punched cards and receiving machine code printed on huge fanfolds from line printers, the ability to type a command on a keyboard and see the computer's immediate response on their own printer was nothing short of miraculous. Through the rapidly spreading use of time-sharing, many people were able to use individual terminals to directly interact with large computers. To those who knew about the plans to connect their time-sharing communities into a supercommunity, 1968 was a time of exciting and rapid change in a field that was still virtually unknown to the outside world.

The idea of a *community* that could be brought into existence by the construction of a new kind of computer system was perhaps the most radical proposal in the 1968 paper. The ARPAnet was not to be officially on-line until 1969, but at that point the time-sharing groups had constructed enough of the superstructure for the outlines of the new network to be known and visible.

Taylor and Licklider were more concerned about the further development of this test-bed for advanced communications and thought amplification than they were dedicated to the use of the network as an operational entity for conducting weapons research. Writing with the knowledge that ARPAnet was to begin operation within a year, and would probably be generally unknown outside defense or computer science circles, Licklider and Taylor pointed out: [4]

> . . . Although more interactive multiaccess computer systems are being delivered now, and although more groups plan to be using these systems within the next year, there are at present perhaps only as few as half a dozen interactive multiaccess computer *communities*.
>
> These communities are socio-technical pioneers, in several ways out ahead of the rest of the computer world: What makes them so? First, some of their members are computer scientists and engineers who understand the concept of man-computer interaction and the technology of interactive multiaccess systems. Second, others of their members are creative people in other fields and disciplines who recognize the usefulness and who sense the impact of interactive multiaccess computing upon their work. Third, the communities have large multiaccess computers and have learned to use them. And fourth, their efforts are regenerative.

The authors were looking beyond the networks of their day, and the computer systems that were commercially available, to the technology they knew would be possible and affordable on a large scale within decades. Convinced that the technology they and their colleagues had created, and the community of users that had grown up around that technology, were but the forerunners to far more powerful and more widely usable systems, they called for the development of a version of certain time-sharing systems into a tool that could be used to amplify human communications: [5]

> . . . These new computer systems we are describing differ from other computer systems advertised with the same labels: interactive, time-sharing, multiaccess. They differ by having a greater degree of open-endedness, by rendering more services, and above all by providing

facilities that foster a working sense of community among their users. The commercially available time-sharing services do not yet offer the power and flexibility of software resources—the "general purposeness"—of the interactive multiaccess systems of the System Development Corporation in Santa Monica, the University of California at Berkeley, Massachusetts Institute of Technology in Cambridge and Lexington, Mass.—which have been collectively serving about a thousand people for several years.

The thousand people include many of the leaders of the ongoing revolution in the computer world. For over a year they have been preparing for the transition to a radically new organization of hardware and software, designed to support many more simultaneous users than the current systems, and to offer them—through new languages, new file-handling systems, and new graphic displays—the fast, smooth interaction required for truly effective man-computer partnership.

Time-sharing, tremendously exciting as it was to programmers, was seen as only a means to an end by those who were aiming to build communication amplifiers. To those who were gung-ho about the future of multiaccess computing, Taylor and Licklider talked about the ultimate goal of the various projects they had initiated: the creation of tools to enhance the thinking of individuals and augment communications among groups of people.

Engelbart's group at SRI, Ivan Sutherland's computer graphics work at MIT and Harvard, David Evans and his students at the University of Utah, the Project MAC hackers at MIT, and other groups scattered around the country were constructing pieces of a whole new technology. Foreseeing the day when such systems would become practical on a large scale, Licklider and Taylor reminded their colleagues that the new information processing technology could revolutionize not only research centers and universities, but offices, factories, and ultimately schools and homes.

Looking toward what was then the long-term future, Licklider and Taylor projected a positive attitude about the possible social impact of supercommunities that might include not only computer scientists and programmers but housewives, schoolkids, office workers, and artists: [6]

But let us be optimistic. What will on-line interactive communities be like? In most fields they will consist of geographically separated members, sometimes grouped in small clusters and sometimes working individually. They will be communities not of common location, but of *common interest*. In each field, the overall community of interest will be large enough to support a comprehensive system of field-oriented programs and data.

In each geographical sector, the total number of users—summed over all the fields of interest—will be large enough to support extensive general-purpose information processing and storage facilities. All of these will be interconnected by telecommunications channels. The whole will constitute a labile network of networks—ever changing in both content and configuration.

The authors envisioned the creation of an interconnected system of software-based tools that would provide "investment guidance, tax counseling, selective dissemination of information in your field of specialization, announcement of cultural, sport, and entertainment events that fit your interests, etc. In the latter group will be dictionaries, encyclopedias, indexes, catalogues, editing programs, teaching programs, testing programs, programming systems, data bases, and—most important—communication, display, and modeling programs." They could have been describing from life the facilities that were available at PARC, ten years later.

Licklider and Taylor were most emphatic that the impact would be great, on both individuals and organizations, when all the elements, which they could only speculate about in 1968, were perfected sometime in the future: [7]

First, life will be happier for the on-line individual because the people with whom one interacts most strongly will be selected more by commonality of interests and goals than by accidents of proximity. Second, communication will be more effective, and therefore more enjoyable. Third, much communication will be with programs and programmed models, which will be (a) highly responsive, (b) supplementary to one's own capabilities, rather than competitive, and (c) capable of representing progressively more complex ideas without necessarily displaying all the levels of their structure at the same time—and which will therefore be both challenging and rewarding. And fourth, there will be plenty of opportunity for everyone (who can afford a console) to find his calling, for the whole world of information, with all its fields and disciplines, will be open to him—with programs ready to guide him or to help him explore.

For the society, the impact will be good or bad, depending mainly on the question: Will "to be on-line" be a privilege or a right? If only a favored segment of the population gets a chance to enjoy the advantage of "intelligence amplification," the network may exaggerate the discontinuity in the spectrum of intellectual opportunity.

On the other hand, if the network idea should prove to do for
education what a few have envisioned in hope, if not in concrete
detailed plan, and if all minds should prove to be responsive, surely
the boon to humankind would be beyond measure.

Strangely lyrical and surprisingly romantic prose coming from two
computer-research organizers in the Pentagon. But by 1971, when Taylor
recruited fifty or sixty of the best people in the field for the Computer
Science Laboratory at PARC, the cream of the interactive computer
designers had enough engineering and software research behind them
from the time-sharing and ARPAnet projects to make them confident
that such a utopian scenario might be possible—especially if a corporation
with the resources of Xerox was willing to take a high-stakes gamble.

The people who built the first interactive, multiaccess computers,
the first intellectual augmentation systems, and the first packet-switching
computer networks were gathering under the same roof for the first
time, in order to turn those dreams into prototypes as soon as possible.
Butler Lampson, Chuck Thacker, Jim Mitchell, Ed McCreight, Bob
Sproull, Jim Morris, Chuck Geschke, Alan Kay, Bob Metcalfe, Peter
Deutsch, Bill English—to those who knew anything about the esoteric
world of computer design, the PARC computer science founders constituted an unprecedented collection of talents.

It wasn't the kind of shop where old-style hierarchies and pecking
orders would do any good. You don't run an outfit like that as much
as you mediate it—which is where Bob Taylor came in. The kind of
thing they were building, and the kind of people it took to build it,
required a balance between vision and pragmatism, the kind of balance
that couldn't be enforced by artificially imposed authority.

What they all agreed upon was what they wanted to get their hands
on, in the way of a first-rate software research facility. The potential
of computers as tools to be used by individuals, and the communications
possibilities opened up by linking computers, were what motivated the
PARC team. It was time to demonstrate that the theories about using
computers to manage personal communications could work in an office
like theirs. If they could demonstrate that such devices could speed
their own work, they would be on the way to selling the rest of the
world on the vision they held from the time-sharing days.

The first thing they needed in order to retool the world of information
work was a computer designed for one person to use, something that
went far beyond previous attempts. Because they knew that vision was
the human sense capable of the most sophisticated informational input,

Bob Taylor and Wes Clark in Taylor's PARC office. The Alto, Ethernet, and other breakthrough prototypes were developed in Taylor's Computer Systems Laboratory. Clark was a designer of the Whirlwind and TX-2 computers at MIT, a designer of the LINC, the first attempt at building a personal computer, and the architect of the "intermediate message processor" idea that made computer networks possible. (Courtesy of Bob Taylor.)

the PARC computerists knew they wanted a sophisticated graphic screen to bring the computer's power out to the user. Complex, dynamic, visual models required a large amount of computer power, so the decision to emphasize the visual display meant that the hardware would have a great deal more memory and speed than anyone else in the computer world had heretofore put at any one individual's command.

"We wanted hardware as capable as we could afford to build," Taylor recalls, "because we needed capable computing tools to design and construct an entire software architecture that nobody in the world yet knew how to make. We wanted for our own use what we thought other information workers would eventually want. We needed the computing power and the research environment to build something expensive but very flexible and growable that would someday be much less expensive but even more capable. We all understood when we planned the Alto that the main memory of what we wanted might cost $7000 by the time it was produced, in 1974, but would drop to around $35 ten years later."

The hardware shop at PARC was only set up to produce small batches for the PARC software designers, but eventually 1500 Altos were built for Xerox executives and researchers, for associates at SAIL and SRI, as well as for the U.S. Senate, House of Representatives, certain other government agencies, and even the White House staff. It was the first machine designed to put significant computing power on a person's desk.

The job the Alto designers did in 1973 was all the more remarkable when compared with the first "personal computers" the outside world was to learn about years later. The 1975 Altair, the granddaddy of the homebrew computers, had all of ¼K main memory (also known as RAM, this represents that amount of storage space the computer devotes to its "working memory," and thus indicates the rough limit of how much work it can do with reasonable speed). The first Apple models sold, in 1977, had 8K. When IBM introduced its personal computer, in 1981, the standard model had 16K. The Alto, in 1974, started at 64K and was soon upgraded to 256K. The distinctive bit-mapped screen and the mouse pointing device weren't to be seen on a non-Xerox product until 1983, when Apple produced Lisa.

The hardware, of course, was just part of the story. These devices were built for the people whose job was to create equally spectacular software innovations. And the personal computers themselves weren't enough for those who longed for the kind of community they had known with the ARPAnet.

"We didn't start talking about the hardware and the software until we talked about what we wanted to do personally with such a system," Taylor remembers. "We knew there were technical problems to solve, and we would challenge them in due time. First we had to consider the human functions we wanted to amplify. For example, people use their eyes a great deal to assimilate information, so we knew we wanted a particularly powerful kind of display screen. Then all the time-sharing veterans insisted they wanted a computer that didn't run faster at night."

What Taylor meant was that the time-sharing programmers had all become accustomed in the mid-1960s to doing their serious computing in the middle of the night, when the amount of traffic on the central computer was light enough to perform truly large information processing tasks without delay. The first radical idea they agreed upon was that each Alto had to have as much main memory as one of the central computers from the time-sharing systems of only a few years back. And it had to be fast.

"People can give commands to a computer much more rapidly and easily by seeing and pointing than by remembering and typing, so we adopted and then adapted the mouse," added Taylor. "It is hard for people to learn artificial languages and even harder for machines to learn natural languages. The existing computer languages didn't give first-time users and experienced programmers equal power to interact with the computer, so we created new kinds of languages."

"Most importantly, people often need to do things in *groups*. There are times when we want to use the Alto as a personal tool, and times when we want to use it as a communication medium, and times when we want both. Our purpose in bringing all that computing power to individuals was not to allow them to isolate themselves. We wanted to provide the gateway to a new information space, and ways to fly around in it, and a medium for community creativity, all at the same time."

When time-sharing first got going, and the hackers began to proliferate late at night in the corners of university computer departments, the subcult of computerists found that while they could all communicate with the central computer at the same time, they couldn't all necessarily communicate with each other, or share each other's programs or files. It took some effort, but the time-sharing systems programmers eventually solved the problem.

The solution to the difficult problem of sharing resources among different users of a multiaccess computer became no less difficult when it had to be translated to the problem of sharing resources between many equally powerful, geographically separated, often incompatible computers (as with ARPAnet). The carefully designed connectivity of time-sharing systems could not be patched onto the new system.

The PARC network had to be built from the ground up, along with the personal workstations and shared *servers* for filing, printing, and mail. The server notion meant that certain otherwise stock-model Altos would be programmed for the tasks of controlling these network services, instead of building separate devices to perform these tasks. The concept of the resulting Ethernet, as it was called, stemmed from the determination to make the network itself a tool at the command of the individual user.

The PARC folks were hungry for personal computing power, but they didn't want to give up that hard-won and effort-amplifying community they were just beginning to come to know on the ARPAnet. Dan Swinehart, an SRI alumnus who joined PARC early in the game, remembers that "From the day the Alto was proposed, Butler Lampson and Bob Metcalfe pointed out that if we were going to give everybody at PARC a self-contained computer instead of hooking them all into a central time-sharing system, we'd need a connecting network with enough communicating and resource-sharing capability that the people at the personal work stations wouldn't be isolated from each other."

Thus, the companion project to the Alto was the Ethernet, the first of the "local area networks." With the advent of network technology, the hardware became less important and the software became more important, because such a network consists of a relatively simple hardware level, where a small box physically plugs the individual computer into the network, and a series of more sophisticated software levels known as *protocols* that enable the different devices to interoperate via a communication channel.

With common-carrier networks—the kind where teenage hackers use their telephones to gain access to Defense Department computers—the small box is known as a *modem* and works by translating computer bits into a pattern of tones that the public telephone system uses to communicate information. A local area network uses a different kind of small box that converts computer data into electrical impulses that travel from computer to computer via a short cable, rather than audio tones that are sent over common-carrier communication lines.

Local area networks are meant for environments like PARC—any campus or laboratory or group of offices where many machines are distributed over a small geographical area. But the range of the network is not necessarily limited to the local area. Several local networks can also be linked over long distances via "message processors" known as *gateways* to the common-carrier-linked internetwork. This scheme embeds local networks in more global supernetworks.

Today's network technologies use the *packet-switching* techniques originally developed during the creation of the ARPAnet—exactly the kind of coding of information that Shannon predicted in 1948. Information is transported and processed in packets of information—bursts of coded on-off pulses—that carry, in addition to the core data of the message, information on how the message is to be transmitted and received. If your computer uses the right kind of hardware and software translators, your data will find its own way through the network according to the control and routing information embedded in the packets.

The technical details of packet switching won't matter to the vast majority of the people who will end up using network systems in the future, but the notion of "distributed computing" signals an important change to a new phase in the evolution of computation. Distributed systems, in which a number of people, each with their own significantly powerful personal computers, join together into even more powerful computational communities, are altogether different from the centrally controlled and highly restricted computers of the early days.

Where we will all choose to go, or be forced to go by human nature or historical circumstances, once we are given access to such a system, is a wide-open question, once you get beyond the revolutionary but relatively simple applications to office work. Almost all the augmentation pioneers now use the analogy of the early days of automobiles to describe the present state of the system. Engelbart and Taylor agree that the personal computers millions of enthusiasts are using today are not even at the stage the automobile industry reached with the Model T. More important, there is not yet a uniform and widespread transportation support structure for the messages between individuals.

There are no standard ways to build or drive the informational vehicles that have been devised only recently. The existing highways for large-scale, high-bandwidth information transportation don't even cover a fraction of the countryside. There are no service stations or road maps. The tire industry and the petroleum industry of the knowledge age don't exist yet. There may be prototypes of mind-extending technologies at places like PARC, but there is not yet an infrastructure to support their use in the wider society.

The researchers at PARC were wildly successful in their efforts to build powerful personal computers, years before the business and consumer communities were prepared to accept them, but Xerox marketing management failed to take advantage of the head start achieved by their research and development teams by quickly turning the prototypes into products. The failure of Xerox to exploit the research breakthroughs at PARC was partially a result of the lack of the kind of infrastructure described by the automobile analogy. Technology transfer in such a fast-moving field as microelectronic devices is a tough enough gamble. The problem gets more complicated when those devices are intended to affect the way people think. Building a system from scratch and showing that it works is still a long way from convincing most of the people in the work force to change the way they've always done things.

By the mid-1970s, the nation's smartest computer researchers realized that the Alto, Ethernet, and Smalltalk (an equally advanced computer language) prototypes created at PARC had advanced the state of interactive computing far beyond the level achieved by the ARPA-sponsored time-sharing projects that had revolutionized computers a decade previously. By the late 1970s, Xerox management was ready to think about turning PARC's successes into a product.

While the PARC whiz kids raced ahead on advanced research into dozens of information-related sciences and technologies, the *Star* and the *Ethernet* were readied for the market. Star was designed to be much more than a production-model Alto: The main memory was 512K, twice as much as the enhanced Alto, and the Star's processor was built to run three times as fast as the Alto. The Star's software included a language named Mesa (created in Taylor's lab), along with a whole toolkit of application programs for editing, filing, calculating, computing, creating graphics, distributing electronic mail.

One of the clichés of the computer industry in the early 1980s was that "if Xerox had marketed the Star when it was technically ready to go, they would have stolen an industry out from under IBM and Apple." As it happened, April, 1981, when the Star 8010 Information System was announced, was still too early for the larger segment of office professionals to realize that they were information workers. Xerox marketing management insisted that the workstation was not only a breakthrough in providing tools for individuals, but a part of an integrated office system of interconnected components that shared mail, printing, and filing services. But nobody outside a few privileged test sites knew what that meant.

Until word processing came out of nowhere (as far as the people in offices were concerned) to replace most of the typing pools in the early 1980s, it wasn't clear to the people who bought office equipment for corporations that computers and office workers were bound to get acquainted rapidly. To the first knowledge workers at aerospace firms, financial companies, and technical publishing shops who tested the Star-Ethernet systems, it was very clear that there was a major difference between these machines and the devices they had formerly known as computers.

The place where the mind meets the machine, the long-neglected frontier of computer development, was advanced to a new level by those at ARC and PARC who created the partially psychological, partially computational engineering of the *user interface*. The dreams of the augmentation pioneers were finally materialized in the products of their students, who took the first steps with the Star to engineer the machine to the minds of the potential users. The Star designers reiterated the connection between sophisticated visual representation and the ability to amplify thought: [8]

> During conscious thought, the brain utilizes several levels of memory, the most important being the "short-term memory." Many studies have analyzed the short-term memory and its role in thinking. Two conclusions stand out: (1) conscious thought deals with concepts in the short-term memory . . . and (2) the capacity of the short-term memory is limited. . . . When everything being dealt with in a computer system is visible, the display screen relieves the load on the short-term memory by acting as a sort of "visual cache." Thinking becomes easier and more productive. A well designed computer system can actually improve the *quality* of your thinking. . . .
>
> A subtle thing happens when everything is visible: *the display becomes reality.* The user model becomes identical with what is on the screen. Objects can be understood purely in terms of their visible characteristics.

The idea that the right kind of computer systems could affect the way people think—the seed planted by Vannevar Bush and nurtured by Licklider and Engelbart—was not lost on the Xerox interface builders. In regard to the principle that they called "consistency," the Star team noted: [9]

> One way to get consistency into a system is to adhere to *paradigms* for operations. By applying a successful way of working in one area to other areas, a system acquires a unity that is both apparent and real. . . .
>
> These paradigms *change the very way you think.* They lead to new habits and models of behavior that are more powerful and productive. They can lead to a *human-machine synergism.*

After ten years, PARC had achieved its technological goals, and more. The Alto and Ethernet projects had been wildly successful. The Mesa and Smalltalk languages were both significant advancements of the software art. If bold and imaginative research were all that the success of a company depended on, Xerox would have been in a position to challenge even the dominating force of the information industry. But Peter McColough was no longer the CEO, and Xerox top management failed to comprehend the ten-year technological lead their research division had handed them.

Some of the most important members of the starting team left PARC in the early 1980s, to join other companies or to start their own firms. Such job changes at the higher levels of the electronics and computer industries were far from unknown in Silicon Valley; in fact, PARC was distinguished from similar institutions for many years because of the unusual lengths of time put in by its principal scientists. But when Xerox failed to become the first name in the industry, and the hobbyist side of personal computing had grown to the point where some of the original hobbyists were recruiting PARC scientists and building their own personal computer empires, the first high-level PARC defectors began to seed the rest of the industry with the user interface concepts embodied in the Star.

Bob Metcalfe, the man responsible for the creation of the Ethernet, left to start 3-Com, a company specializing in local area network technology. Alan Kay, whose Smalltalk team made impressive contributions to the Star interface, left to become chief scientist at Atari. John Ellenby, who helped reengineer the Alto 2, became the chairman of Grid. In the fall of 1983, Bob Taylor resigned, after thirteen years leading the laboratory team he had built.

Several of the PARC alumni became associated with those industry newcomers who had emerged from the homebrew computer days. Some of the former whiz kids from PARC were making alliances with the next generation of whiz kids. Charles Simonyi, by then in his early thirties, who was in charge of producing the word processing software for the Alto, left PARC to join Bill Gates, the twenty-seven-year-old chairman of Microsoft, a company that started out as a software supplier to the computer hobbyists in the Altair days of 1975, and is now the second-largest microcomputer software company in the world.

Steve Jobs, chairman of Apple, then in his late twenties, visited PARC in 1979. He was given a demonstration of the Alto. Larry Tesler, the member of the PARC team who gave Jobs that demonstration, left PARC in 1980 and joined Apple's new secret project that Jobs promised would redefine the state of the art in personal computers. In 1983 Apple unveiled Lisa—a machine that used a mouse, a bit-mapped screen, windows, and other features based on the Star-Alto-Smalltalk user interface. The price for the system was around $10,000. This was $6000 less than the more powerful Star, but still hardly in the range of the consumer market. In 1984, Apple brought out a scaled-down, cheaper version of Lisa, the Macintosh, with the same user interface, and revolutionized the personal computer market.

If time-sharing research had been the unofficial initiation ceremony and the ARPAnet was the rite of passage, the PARC era was the end

of the apprenticeship era for the augmentation community. New generations of researchers and entrepreneurs were entering the software fray
through the infant personal computer industry. By the early 1980s, it
didn't take a computer prophet to see that big changes were going to
continue to happen as the mass market began to awaken to the potential
of personal computing. Although the hardware and software of the first
tens of millions of personal computers fell far short of what the PARC
veterans were working toward, the stakes of the game had changed with
the emergence of a mass market.

The beginnings of a much wider computer-using community also
meant the end of arcane jargon and software designs that required complex interactions with the computer. The design principles demonstrated
by the Star and the Lisa pointed the way for future computer designers.
At PARC, they were already onto the Dorado, the Dolphin, and other
post-Star computers. Now that truly capable computing machinery was
becoming available, it was becoming more widely known that the commercially successful programs of the future would be those that succeeded
in bringing the power of the computer out to the person who needs
to use it.

The "rule of two" is, incredibly, still in effect, promising even more
powerful and less expensive computer hardware in the late 1980s. In
1984, Bob Taylor, now with Digital Equipment Corporation, started
doing what he does best—assembling a computer systems research team
for a final assault on the objective. Some of the key members of his
team were graduate students when ARPA funded time-sharing, and had
been involved in the ARC and PARC eras. The latest arena for their
ongoing effort to bootstrap interactive computation technology to the
threshold of truly powerful personal computing was named "Systems
Research Center"—or SRC, pronounced "circ" ("as in circus").

"Come to my office in five years," Taylor challenged me, at the beginning of this gun-lap in the augmentation quest, "and I'll show you a
desktop machine twice as fast as the biggest, most expensive supercomputer made today. Then it will become possible to create the software
that can take advantage of those capabilities we've known about for a
long time."

Taylor now believes that three factors will lead to the most astonishing
plateau in information processing we've seen yet: first, a new level of
systems software will be able to take advantage of computer designs
that make each personal workstation into a kind of miniature distributed
network, with multiple parallel processors inside working in coordination;
second, large scale integration processors will be small enough and cheap

enough to put fast, vast memory into desktop machines; third, and most important, the people who built time-sharing, graphics, networks, personal computers, intelligent user interfaces, and distributed computing are now at the height of their powers, and they have put hundreds of thousands of person-hours into learning how to build new levels of computer technology.

Advances in network technologies, graphics, programming languages, user interfaces, and cheap, large-scale information storage media mean that the basic capabilities dreamed of by the designers of the first personal computers are likely to become widely available before the turn of the century. One hopes that we will be ready to use them wisely. It would be a sad irony if we were to end up creating a world too complicated for us to manage alone, and fail to recognize that some of our own inventions could help us deal with our own complexity.

CHAPTER ELEVEN

The Birth of the Fantasy Amplifier

When millions of portable, affordable, imagination amplifiers fall into the hands of eight-year-old children, look for Alan Kay somewhere in the plot. He has always been too impatient to wait for someone else to bring him what he wanted. And he's always found ways to create what he wanted if it didn't exist. For the past fifteen years, his sights have been set on handheld, full-color, stereophonic, artificially intelligent, information representation toys. And he wants them by the tens of millions. They don't exist yet, so he's enlisted some formidable allies to help him create them.

Fame, fortune, or even the more esoteric career ambitions of top-notch software professionals do not seem to motivate Dr. Kay, now a "research fellow" for Apple, formerly "chief scientist" at Atari Corporation. Becoming another Silicon Valley millionaire or accepting the offer of an endowed chair at MIT have not interested him as much as the prospect of putting the power to *imagine* into the hands of every bright kid who ever got thrown out of a classroom.

Ever since he learned to read at the age of two and a half, Alan Kay has been accustomed to doing things his own way and letting the rest of the world catch up later. At the same time that he was close to flunking out of the eighth grade, primarily for insubordination, he was one of television's original "Quiz Kids." Ten years before he coined the term "personal computer," before Atari or PARC existed, and before another pair of bright insubordinates named Wozniak and Jobs created a new meaning for that good old American word "Apple," Alan Kay was demonstrating FLEX, a personal computer in all but name, to the ARPA graduate students' conference.

Alan is now in his early forties, and is acknowledged by his peers, if not yet the general public, as one of the contemporary prophets of the personal computer revolution. Now his goal is to build a "fantasy amplifier," a "dynamic medium for creative thought" that is powerful enough, small enough, easy enough to use, and inexpensive enough for every schoolkid in the world to have one. He has the resources and the track record to make you believe he'll do it.

Alan Kay doesn't fit the popular image of the arrogant, antisocial hacker, the fast-lane nouveau micromillionaire, or the ivory tower computer scientist. He wears running shoes and corduroys. He has a small, meticulous moustache and short, slightly tousled dark hair. He's so image-less you could pass him in the halls of the place he works and not notice him, even though he's the boss. Which isn't to say that he's egoless or even modest. He loves to quote himself, and often prefaces his homilies with phrases like "Kay's law number one states . . . ."

When I first encountered him, between his stint as director of the legendary "Learning Research Group" at Xerox PARC, and his present position as a kind of "visionary at large" for Apple, Dr. Kay and his handpicked team at Atari were working under tight secrecy, with a budget that was rumored to be somewhere between $50 million and $100 million, to produce something that nobody in the corporation ever described to anybody outside the corporation. But anybody who has ever talked to him, or read something he has written about his dreams, can guess the general thrust of Kay's Atari project, and the probable direction of his current work at Apple. He has been moving toward realizing his dream, project by project, prototype by prototype, innovation by innovation, ever since he was a graduate student.

Being the kind of person he is didn't make it easy for Alan to get an education. At the beginning, he knew more than all of his classmates and most of his teachers, and he didn't mind demonstrating it aloud— a trait that got him thrown out of classrooms and beaten up on playgrounds.

Fortunately for him and for all of us who may benefit from his creations in the future, Alan was already well armored in his mind and imagination, where it really counted, by the time his teachers and classmates got ahold of him. For Alan, being ahead of everybody else started out as a pleasure and quickly turned into a survival trait—which meant he didn't do too well in school, or anyplace else, until the summer of his fifteenth year, when "a music camp in Oneonta, New York, changed my entire life."

Music became the center of his life. In many ways, it still is. He commutes to Silicon Valley from his home in Brentwood, 300 miles away, mostly because he doesn't want to be away from his homemade pipe organ for too long. And he still goes to music camp every summer. He never could understand why his two favorite toys—books and musical instruments—could not be combined into a single medium capable of dealing with both sounds and symbols. He worked as a professional jazz and rock guitarist for ten years. When it looked like he was about to be drafted, Kay joined the U.S. Air Force as a navigational cadet. In the Air Force, he "wore out a pair of shoes doing insubordination duty," but he also learned that he had a knack for computer programming.

After he finished his Air Force duty, the National Center for Atmospheric Research was eager enough to use Kay's programming talent to pay his way through the University of Colorado. He earned a degree in biology, but his college grades were as mixed as they had always been, because of his habit of concentrating intensely on only those things that interested him. Through what Alan now calls "sheer luck," he came to the attention of somebody smart enough to actually teach something to a smartass like Alan Kay—and bold enough to admit a student with an undergraduate record that read more like a rap sheet than a transcript.

The man who gambled on Kay's checkered history in academia was David Evans, the chairman of the computer science department at the University of Utah, a place that was to become one of the centers of the universe for the augmentation community by the mid-1960s. Like so many others who assumed positions of leadership in the field of interactive computer systems design, Evans had been involved in early commercial computer research and with the ARPA-funded groups that created time-sharing.

"These career pathways of ARPA project leaders and their graduate students repeatedly intertwined," Kay recalls. "An enormous amount of work was done by a few people who kept reappearing in different places over the years. People frequently migrated back and forth from one ARPA project or another. ARPA funded people rather than projects, and they didn't meddle for an extended period. Part of the genius of

Alan Kay, 1984. (Courtesy of Apple Computer, Inc.)

Licklider and Bob Sproull was the way this moving around contributed to the growth of a community."

One of the people Evans managed to recruit for the Utah department who had an impact, not only on Alan Kay but on the entire course of personal computing, was Ivan Sutherland, the graduate student and protégé of Claude Shannon and J.C.R. Licklider who single-handedly created the field of computer graphics as part of his MIT Ph.D. thesis—the now-legendary program known as "Sketchpad."

People like Alan Kay still get excited when they talk about Sketchpad: "Sketchpad had wonderful aspects, besides the fact that it was the first real computer graphics program. It was not just a tool to draw things. It was a program that obeyed laws that you wanted to be held true. So to draw a square in Sketchpad, you drew a line with a lightpen and said: 'Copy-copy-copy, attach-attach-attach. That angle is 90 degrees, these four things are to be equal.' Sketchpad would go zap! and you'd have a square."

Another computer prophet who saw the implications of Sketchpad and other heretofore esoteric wonders of personal computing was an irreverent, unorthodox, counterculture fellow by the name of Ted Nelson, who has long been in the habit of self-publishing quirky, cranky, amazingly accurate commentaries on the future of computing. In *The Home Computer Revolution*[1] Nelson had this to say about Sutherland's pioneering program, in a chapter entitled "The Most Important Computer Program Ever Written":

> You could draw a picture on the screen with the lightpen—and then file the picture away in the computer's memory. You could, indeed, save numerous pictures in this way.
>
> You could then combine the pictures, pulling out copies from memory and putting them amongst one another.
>
> For example, you could make a picture of a rabbit and a picture of a rocket, and then put little rabbits all over a large rocket. Or, little rockets all over a large rabbit.
>
> The screen on which the picture appeared did not necessarily show all the details; the important thing was that the details were *in* the computer; when you magnified a picture sufficiently, they would come into view.
>
> You could magnify and shrink the picture to a spectacular degree. You could fill a rocket picture with rabbit pictures, then shrink that until all that was visible was a tiny rocket; then you could make copies of *that*, and dot them all over a large copy of the rabbit picture. So when you expanded the big rabbit till only a small part showed (so it would be the size of a house, if the screen were large enough), then the foot-long rockets on the screen would each have rabbits the size of a dime.
>
> Finally, if you changed the master picture—say, by putting a third ear on the big rabbit—all the copies would change correspondingly.
>
> Thus Sketchpad let you try things out before deciding. Instead of making you position a line in one specific way, it was set up to allow you to try a number of different positions and arrangements, with the ease of moving cut-outs around on a table.
>
> It allowed room for human vagueness and judgement. Instead of forcing the user to divide things into sharp categories, or requiring the data to be precise from the beginning—all those stiff restrictions people say "the computer requires"—it let you slide things around to your heart's content. You could rearrange till you got what you wanted, no matter for what reason you wanted it.
>
> There had been lightpens and graphical computer screens before, used in the military, but Sketchpad was historic in its simplicity—a simplicity, it must be added, that had been deliberately crafted by a cunning intellect—and its lack of involvement with any particular field. Indeed, it lacked any complications normally tangled with what people actually do. It was, in short, an innocent program, showing how easy human work could be if a computer were set up to be really helpful.
>
> As described here, this may not seem very useful, and that has

been part of the problem. Sketchpad was a very imaginative, novel program, in which Sutherland invented a lot of new techniques; and it takes imaginative people to see its meaning.

Admittedly the rabbits and rockets are a frivolous example, suited only to a science-fiction convention at Easter. But many other applications are obvious: this would do so much for blueprints, or electronic diagrams, or all the other areas where large and precise drafting is needed. Not that drawings of rabbits, or even drawings of transistors, mean the millennium; but that a new way of working and seeing was possible.

The techniques of the computer screen are general and applicable to *every*thing—but only if you can adapt your mind to thinking in terms of computer screens.

Sutherland was twenty-six when he succeeded Licklider as director of ARPA's Information Processing Techniques Office. Then he was succeeded by Bob Taylor when he left for Harvard in the mid-1960s, to work on 3-D head-mounted displays (like miniature televisions in eyeglass frames) and other exotic graphics systems. When David Evans tried to lure him to Utah, Sutherland said he would do it if Evans agreed to become a business partner—and thus the pioneering computerized flight simulation and image generation company of Evans & Sutherland was born.

Kay showed up at Utah in November of 1966. His first task was to read a pile of manuscript Evans gave him—Ivan Sutherland's thesis. The way Evans ran the graduate program, you weren't supposed to be around campus very long or very much. You were supposed to be a professional and move on to high-level consulting jobs in industry. The job he found for Alan Kay was with a hardware genius named Ed Cheadle. Ed had an idea about doing a tabletop computer. Kay worked on FLEX— his first personal computer software design—from 1967 to 1969. While some of the founders of today's personal computer industry were still in high school, Kay was learning how to design personal computers.

Technically, Cheadle and Kay were not the first to attempt to build a personal computer. Wes Clark, from Whirlwind and Lincoln Lab's TX-2 and "imps," had constructed a desk-size machine a few years before, known as "LINC." FLEX was an attempt to use the more advanced electronic components that had recently become available to bring more of the computer's power out where the individual user could interact with it. FLEX was a significant innovation technically, but it was complicated and delicate and, in Kay's words, "users found it repellent to learn." The problem wasn't in the machinery as much as it was in the special language the user had to master in order to command the power of the machine to accomplish useful tasks. That was when Kay first vowed to make sure his personal computer would come at least part of the way toward the person who was to use it, and when he realized that software design would be the area in which this desire could be fulfilled.

Although he didn't fully realize it yet, Alan Kay was beginning to think about designing a new programming language. The kind of language he began to yearn for would be a tool for using the computer as a kind of universal simulator. The problem was that programming languages were demonically esoteric. "There are two ways to think about building an instrument," Kay asserts. "You can build something like a violin that only a few talented artists can play. Or you can make something like a pencil that can be used quickly and easily for anything from learning the alphabet to drawing to writing a computer program." He was convinced that 99 percent of the problems to be solved in making a truly usable personal computer were software problems: "By 1966, everyone knew where the silicon was going."

Besides FLEX, Kay's other project at Utah was to make some software work. He got a pile of tape canisters on his desk, along with a note saying that the tapes were supposed to contain a scientific programming language known as Algol 60, but they didn't work. It was a maddening software puzzle that was still far from solved when Kay figured out that it wasn't Algol 60 at all, but a language from Norway, of all places, called *Simula*. In a 1984 interview, Kay described what happened when he finally printed out on paper the program listings stored in those mysterious canisters and figured out what was on those tapes: [2]

> We couldn't understand any of the papers, they were sort of trans-literated from the Norwegian. . . . We spread out the program listings and actually went through the machine code to try to figure out what was happening—and I suddenly realized that Simula was a pro-gramming language to do what Sketchpad did. I had never really understood what Sketchpad was. I get shivers now thinking of it. It rotated my point of view through a different dimension and nothing has been the same since. I suddenly understood the purpose of higher-level languages.

Alan was one of the enthralled audience at Engelbart's 1968 media show. He was excited by it because it demonstrated what you could really do with a computer-augmented representation system. It also made it clear to Alan what he *didn't* want to do. "The Engelbart crew were all ace pilots of their NLS system," Kay remembers. "They had almost instant response—like a very good video game. You could pilot your way through immense fields of information. It was, unfortunately for my purposes, something elegant and elaborate that these experts had learned how to play. It was too complex for my tastes, and I wasn't interested in the whole notion of literacy as a kind of fluency."

In the course of preparing his Ph.D. thesis, Alan began to explore the world of artificial intelligence research, which brought him into closer contact with two more computer scientists who were to heavily influence his own research—Marvin Minsky and Seymour Papert, who were then codirectors of MIT's pioneering artificial intelligence research project. In the late 1960s, Papert in particular was doing something that irrevocably influenced Alan's goals. Papert was creating a new computer language. For children.

Papert, a mathematician and one of the early heroes of the myth-shrouded Project MAC, had spent five years in Switzerland, working with the developmental psychologist Jean Piaget. Piaget had triggered his own revolution in learning theory by spending his time—years and decades—watching how children learn. He concluded that learning is not simply something adults impose upon their offspring through teachers and classrooms, but is a deep part of the way children are innately equipped to react to the world, and that all children construct their notions of how the world works, from the material available to them, in definite stages.

Piaget was especially interested in how different kinds of *knowledge* are acquired by children, and concluded that children are *scientists*—they perform experiments, formulate theories, and test their theories with more experiments. To the rest of us, this process is known as "playing," but to children it is a vital form of research.

Papert recognized that the responsiveness and representational capacity of computers might allow children to conduct their research on a scale never possible in a sandbox or on a blackboard. LOGO, the computer language developed by Papert, his colleague Wallace Feurzeig, and others at MIT and at the consulting firm of Bolt, Beranek & Newman, was created for a purpose that was shockingly different from the purposes that had motivated the creation of previous computer languages. FORTRAN made it easier for scientists to program computers. COBOL made it easier for accountants to program computers. LISP, some might say, made it easier for computers to program computers. LOGO, however, was an effort to make it easier for *children* to program computers.

Although its creators knew that the LOGO experiment could have profound implications in artificial intelligence and computer science as well as in education, the project was primarily intended to create a tool for teaching thinking and problem-solving skills to children. The intention was to empower rather than suppress children's natural desire to solve problems in ways they find fun and rewarding. "The object is not for the computer to program the student, but for the student to program the computer," was the way the LOGO group put it.

Beginning in 1968, children between the ages of eight and twelve were introduced to programming through the use of attractive graphics and a new approach that put the power to learn in the hands of the people who were doing the learning. By learning how to use LOGO to have fun with computers, students were automatically practicing skills that would generalize to other parts of their lives.

Papert had observed from both his computer science and developmental psychology experience that certain of these skills are "powerful ideas" that can be used at any age, in any subject area, because they have to do with *knowing how to learn*. This is the key element that separated LOGO from the "computer-assisted instruction" projects that had preceded it. Instead of treating education as a task of transferring knowledge from the teacher to the student, the LOGO approach was to help students strengthen their ability to discover knowledge on their own.

One of the most important of these skills, for example, is the idea of "bugs"—the word that programmers use to describe the small mistakes that inevitably crop up in computer programs, and which must be tracked down before the program will work. Instead of launching students on an ego-bruising search for the "right" answer, the task of learning LOGO was meant to encourage children to solve problems by daring to try new procedures, then debugging these procedures until they work.

The first revolutionary learning instrument introduced in LOGO was the "turtle," a device that is part machine and part metaphor. The original LOGO turtle was a small robot, controlled by the computer and programmed by the child, that could be instructed to move around, pulling a pen as it moved, drawing intriguing patterns on paper in the process. Alan Kay was one of several software designers who realized that this process was more than just practice at drawing pictures, for the ability to manipulate symbols—whether the symbols are turtle drawings, words, or mathematical equations—is central to every medium used to augment human thinking.

The abstract turtle of today's more advanced display technology is a triangular graphic figure that leaves a video trail behind it on a display screen. Whether it is made of metal and draws on paper, or made of electrons and draws on a video screen, the turtle is what educational psychologists call a *transitional object*—and what Papert calls an "object-to-think-with."

Instead of "programming the computer" to draw a pattern, children are encouraged to "teach the turtle" how to draw it. They start by "pretending to be the turtle" and trying to guess what the turtle would do in order to trace a square, a triangle, a circle, or a spiral. Then they give the turtle a series of English-like commands, typed in through a keyboard, to "teach the turtle a new word."

If the procedure followed by the turtle in response to the typed commands doesn't achieve the desired graphic effect, the next step is to systematically track down the "bug" that is preventing success. The fear of being wrong is replaced in this process by the immediate feedback of discovering powerful ideas on one's own.

After decades of research, Papert summarized the results of his LOGO work for a general audience in *Mindstorms: Children, Computers, and Powerful Ideas*. In this manifesto of what has grown into an international movement in both the educational and computing communities, Papert reiterated something important that is easy to lose in the complexities of the underlying technology—that the purpose of any tool ought to be to help human beings become more human: [3]

> In my vision the computer acts as a transitional object to mediate relationships that are ultimately between person and person. . . .
> I am talking about a revolution in ideas that is no more reducible to technologies than physics and molecular biology are reducible to the technological tools used in the laboratories or poetry to the printing press. In my vision, technology has two roles. One is heuristic: The computer presence has catalyzed the emergence of ideas. The other is instrumental: The computer will carry ideas into a world larger than the research centers where they have incubated up to now.

When he came across the LOGO work, during the time he was meditating about the fact that he had put two years into the FLEX machine only to find that it wasn't amenable to humans who tried to use it, Alan Kay recalls that "it was like a light going on in my head. I knew I would never design another program that was not set up for children."

One of the first things he understood was that a program or a programming language that can be learned by children doesn't have to be a

Seymour Papert, codirector of Project MAC and creator of the LOGO computer language, pictured here with the mechanical "turtle." (Courtesy of the MIT Museum.)

"toy." The toy can also serve as a tool. But that transformation doesn't happen naturally—it comes about through a great deal of work by the person who designs the language. Kay already knew that the most important tools for creating personal computing were to be found in software, but it now dawned on him that the power those tools would amplify would be the power to *learn*—whether the user is a child, a computer systems designer, or an artificial intelligence program.

Although he knew he had a monstrous software task ahead of him if he was to create a means by which even children could use computers as a simulation tool, his FLEX experience and his exposure to LOGO convinced Kay that there was far more to it than just building an easy-to-operate computer and creating a new kind of computer language. It was something akin to the problem of building a tool that a child could use to build a sandcastle, but which would be equally useful to architects who wanted to erect skyscrapers. What he had in mind was an altogether new kind of artifact: If he ended up with something an eight-year-old could carry in one hand and use to create and communicate music, words, pictures, and to consult museums and libraries, would the device be perceived as a *tool* or as a *toy?*

Kay began to understand that what he wanted to create was an entirely new *medium*—a medium that would be fundamentally different from all the previous static media of history. This was going to be the first *dynamic* medium—a means of representing, communicating, and animating thoughts, dreams, and fantasies as well as words, images, and sounds. He recognized the power of Engelbart's system as a toolkit for knowledge workers like editors and architects, scientists, stockbrokers, attorneys, designers, engineers, and legislators. Information experts desperately needed tools like NLS. But Kay was after a more universal, perhaps more profound power.

One of the concepts that played a big part in Papert's LOGO project, and thus influenced Alan Kay and others, was derived from the thinking of John Dewey, whose work encouraged generations of progressive educators. Dewey developed a theory that Piaget later elaborated—that the imaginative play often mistakenly judged by adults to be "aimless" is actually a potent tool for learning about the world. Kay wanted to link the natural desire to explore fantasies with the innate ability to learn from experimentation, and he knew that the computer's power to simulate anything that could be clearly described was one key to making that connection.

Alan wanted to create a medium that was a *fantasy amplifier* as well as an intellectual augmentor. First, he had to devise a language more suited to his purposes than LOGO, a "new kind of programming system that would attempt to combine simplicity and ease of access with a qualitative improvement in expert-level adult programming." With the right kind of programming language, used in conjunction with the high-powered computer hardware he foresaw for the near future, Kay thought that an entirely new kind of computer—a *personal* computer—might be possible.

Such a software advance as the kind Kay envisioned could only be accomplished by using hardware that didn't exist yet in 1969, since the computing power required for each individual unit would have to be several hundred times that of the most sophisticated time-sharing computers of the 1960s. But at the end of the 1960s, such previously undreamed-of computing power seemed to be possible, if not imminent. The year 1969 was pivotal in the evolution of personal computing, as well as in Alan Kay's career. It was the year that the ARPAnet time-sharing communities began to discover that they were all plugged into a new kind of social-informational entity, and enthusiastically began to use their new medium to design the next generations of hardware and software.

After he finished his thesis on FLEX, Kay began to pursue his goal of designing a new computer language in one of the few places that had the hardware, the software, and the critical mass of brain power to support his future plans—the Stanford Artificial Intelligence Laboratory. He had a lot to think about. There were many great programmers, but very few great creators of programming languages.

The programming language for the eventual successor to FLEX was his primary interest, not only because he knew the hardware would be catching up to him, but because he knew that programming languages influence the minds of the people who use computers. In 1977, after the task of creating his new programming language, *Smalltalk*, was accomplished, Kay described the importance of this connection between a programming language and the thinking of the person who uses it: [4]

> The particular structure of a symbolic language is important because it provides a context in which some concepts are easier to think about than others. For example, mathematical notation first arose to abbreviate concepts that could be expressed only as ungainly circumlocutions in natural language. Gradually it was realized that the form of an expression could be of great help in the conception and manipulation of the meaning for which the expression stood. . . .
> The computer created new needs for language by inverting the

traditional process of scientific investigation. It made new universes available that could be shaped by theories to produce simulated phenomena.

The "inverting" of "the traditional process of scientific investigation" noted by Kay was the source of the computer's power of *simulation*. And the ability to simulate ideas in visible form was exactly what a new programming language needed to include in order to use a computer as an imagination amplifier. If Piaget was correct and children are both experimental scientists and epistemologists, a tool for simulating scientific investigation could have great impact on how much and how fast young children and adult computer programmers are able to learn.

According to the rules of scientific induction, first set down by Francis Bacon three hundred years ago, scientific knowledge and the power granted by that knowledge are created by first observing nature, noting patterns and relationships that emerge from those direct observations, then creating a theory to explain the observations. With the creation of a machine that "obeyed laws you wanted to be held true," it became possible to specify the laws governing a world that doesn't exist, then observe the representation created by the computer on the basis of those laws.

Papert called these simulated universes "microworlds," and used LOGO-created microworlds to teach logic, geometry, calculus, and problem-solving to ten-year-olds. Part of the fascination of a good video game lies in the visual impact of its microworld representation and the amount of power given to the player to react to it and thus learn how to control it. In Smalltalk, every object was meant to be a microworld.

Computer scientists talk about *computational metaphors* in computer languages—alternative frameworks for thinking about what programming really does. The most widespread and oldest metaphor is that of a recipe, the kind of recipe you would create for a very stupid but very obedient servant—a list of definite, step-by-step instructions that could produce a desired result when carried out by a mindless instruction-following mechanism. The sequence of instructions is an accurate but limiting metaphor for how a computer operates. It is a reflection of the fact that early computers were built to do just one thing at a time, but to do it very fast and get on to the next instruction.

This model, however, is not as well suited to computers of the future, which will perform many processes at the same time (in the kind of computation that is called parallel processing). Languages based on the dominant metaphors of numerical, serial procedures are much better suited for linear processes like arithmetic and less well suited for exactly those tasks that computers need to perform if they are to serve as representational media. Parallel processing is also a better model of the way human brains handle information.

Starting from concepts set forth in LOGO and in Simula, Kay began to devise a new metaphor in which the string of one-at-a-time instructions is replaced by a multidimensional environment occupied by objects that communicate by sending one another messages. In effect, he started out to build a computer language that would enable the programmer to look at the host computer not as a serial instruction follower, but as thousands of independent computers, each one able to command the power of the whole machine.

In 1969 and 1970, the growing impact of the Vietnam war and the pressure by congressional critics of what they interpreted as "frivolous research" contributed to the death of the "ARPA spirit" that had led to the creation of time-sharing and computer networks. The "Mansfield Amendment" in 1970 required ARPA to fund only projects with immediately obvious defense applications. Taylor was gone. The AI laboratories and the computer systems designers found funding from other agencies, but the central community that had grown up in the sixties began to fragment.

The momentum of the interactive approach to computing had built up such intensity in its small following by the late 1960s that everybody knew this fragmentation could only be a temporary situation. But nobody was sure where, or how, the regrouping would take place. Around 1971, Alan began to notice that the very best minds among his old friends from ARPA projects were showing up at a new institution a little more than a mile away from his office at the Stanford AI laboratory.

By the beginning of 1971, Alan Kay was a Xerox consultant, then a full-time member of the founding team at the Palo Alto Research Center. By this time, the hardware revolution had achieved another level of miniaturization, with the advent of integrated circuitry and the invention of the microprocessor. Xerox had the facilities to design and produce small quantities of state-of-the-art microelectronic hardware, which allowed the computer designers unheard-of power to get their designs up and running quickly. It was precisely the kind of environment in which a true personal computer might move from dream to design stage. Alan Kay was already thinking about a special kind of very powerful and portable personal computer that he later came to call "the Dynabook."

Everybody, from the programmers in the "software factory" who designed the software operating system and programming tools, to the hardware engineers of the Alto prototype computers, to the Ethernet local-area-network team who worked to link the units, was motivated by the burning desire to get a working personal computer in their own hands as soon as possible. In 1971, Alan wrote and thought about something that wasn't yet called a Dynabook but looked very much like it. Kay's Learning Research Group, including Adele Goldberg, Dan Ingalls, and others, began to create Smalltalk, the programming "environment" that would breathe computational life into the hardware, once the hardware wizards downstairs cooked up a small network of prototype personal computers.

One of the most important features of the anticipated hardware was the visual resolution of the display screen. One of the things Alan had noticed when watching children learn LOGO was that kids are very demanding computer users, especially in terms of having a high-resolution, colorful, dynamic display. They were accustomed to cartoons on television and 70-mm wide-screen movies, not the fuzzy images then to be found on computer displays. Kay and his colleagues knew that hardware breakthroughs of the near future would make it possible to combine the interactive properties of a graphic language like Sketchpad with very high-resolution images.

The amount of image resolution possible on a video display screen depends on how many picture elements are represented on the screen. Kay felt that the threshold number of picture elements needed to most strongly attract and hold the attention of a large population of computer users, and give the users significant power to control the computer, would be around one million dots. (The resolution of a standard snapshot is equivalent to about four million dots.) The Alto computer being constructed for PARC researchers—which the Learning Research Group called "an interim Dynabook"—would have around half a million dots.

The technique by which the Alto would achieve its high-resolution screen was called "bit-mapping," a term that meant that each picture element, each dot of light on the display screen, was connected to one bit of information in a specific place in the computer's memory, thus creating a kind of two-way informational map of the screen. If, for example, a specific bit in the computer's "memory map" was turned off, there would not be a dot of light at the location on the screen. Conversely, an "on" bit at a coordinate in the memory map would produce a dot of light at the designated screen location. By turning on and off parts of the bit map through software commands, recognizable graphic images can be created (and changed) on the screen.

Bit-mapping was a major step toward creating a computer that an individual could use comfortably, whether the user is an expert programmer or a beginner. The importance of a visual display that is connected directly to the computer's memory is related to the human talent for recognizing very subtle visual patterns in large fields of information—undoubtedly a survival trait that evolved way back when our ancestors climbed trees and prowled savannas.

Human information processors have a very small short-term memory,

A close-up of a computer screen displaying Smalltalk capabilities. Note the overlapping "windows," the high-resolution graphics, and the mixture of text and graphics. (Courtesy of Xerox Corporation.)

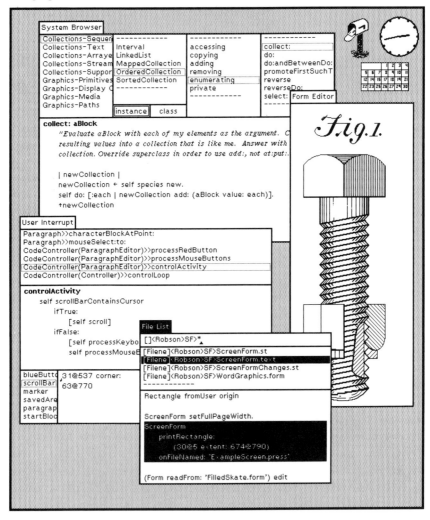

however, which means that while all computers and no humans can extract the square roots of thousand-digit numbers in less than a second, no computers and all humans can recognize a familiar face in a crowd. By connecting part of the computer's internal processes to a visible symbolic representation, bit-mapping puts the most sophisticated part of the human information processor in closer contact with the most sophisticated part of the mechanical information processor.

Bit-mapping created more than a passive window on the computer's internal processes. Just as the computer could tell the human who used it certain facts about whatever it had in its memory, the user was also given the power to change the computer by manipulating the display. If users change the form of the visual representations on bit-mapped screens, using keyboard commands, lightpens (à la Sketchpad), or pointing devices like mice (à la Engelbart), they can also change the computer's memory. The screen is a representation, but it is also a control panel— a drawing on a bit-mapped screen can be nothing more than a drawing, but it can also be a kind of command, even a program, to control the computer's operations.

If, for example, you were to use a mouse to move a video pointer on the screen to touch a visual representation of a file folder or an out basket, and you could call the folder, for example, from the computer's memory and display a document from it on your screen simply by pointing to it, or send the contents of the computer-stored out basket to somebody else's in basket, then a person would be able to use a computer to accomplish the kind of work done in offices, even if that person knew nothing about computer programming. Which, after all, was the potential future market that motivated Xerox management to create PARC and cut their whiz kids loose in the first place.

Creating new kinds of computer input and output devices to help human pattern recognition mesh with mechanical symbol manipulation is known as "designing the human interface," an art and science that had to be created in the 1970s if the kind of human-computer partnership envisioned by Licklider and Engelbart in the 1960s could start to happen by the 1980s. Alan Kay's Smalltalk project played a key role in the evolution of the Alto user interface, and as such was integral to the eventual company goals in the office automation market. But even at the beginning, Kay started bringing children into the project.

Part of the Smalltalk project's effect on the early days at PARC was inspirational. It wasn't long before the rest of the team understood Alan's desire to bring children into the process of designing the same instrument that he and all the other computer scientists wanted to use themselves.

Another aspect of Kay's contribution was more concrete: the absolute conviction that they were designing something meant for people to use. That might not sound too revolutionary today, but even as late as 1971, most of the top-flight computer scientists who believed that this tool was going to be more than just a gadget for computer programmers were at PARC.

PARC in the early 1970s was a collection of the world's best computer scientists, hardware engineers, physicists, programmers . . . which meant it was also a collection of people with strong personalities and definite opinions. Bob Taylor, Alan Kay, Butler Lampson, Bob Metcalfe, and their colleagues each had his own unique approach to creating personal computing, but they agreed on one fundamental assumption—that their ultimate product should be as generally useful as a hammer, or pulley, or book. Secretaries and business executives would one day be able to use the same tool to help them perform office work. Architects and designers would have the power of modeling, forecasting, and simulation at their fingertips. A true personal computer, the diverse PARC groups agreed, ought to be usable by legislators and librarians, teachers and children. And a computer that could be commanded by looking at images on a screen and pointing to them by means of a mouse was certainly a lot more widely usable than a computer that required arcane keyboard-entered commands in order to function.

The first Alto personal computer prototypes were distributed to PARC researchers in 1974. As they had predicted, the creation of an environment in which every researcher had, for the first time in history, personal access to a powerful computer, and the means to communicate with all of his or her colleagues' computers, had a profound effect on their ability to do their job of designing even more powerful computer systems.

By the late 1970s, yet another generation of even more advanced hardware and software had been created by a network of nearly a thousand researchers at PARC equipped with Altos, communicating via Ethernet networks. But the outside world, and many people in the computer world, were still unaware of the potential of personal computers. The problem, as PARC alumnus Charles Simonyi was to point out in 1983, an eventful decade later, was that Xerox management couldn't be faulted for not realizing in 1973 that PARC was more than ten years ahead of an industry that wouldn't even exist until 1975.

Another small cloud on the horizon in the mid-1970s—the "home-brew" computer hobbyists who were building their own low-power micro-computers—became a gathering storm of popular interest in personal computing by the end of the 1970s. The microcomputer hobbyists, who assembled the new microprocessor chips into operational computers, were for the most part unaware of the far more powerful devices that were in use in Palo Alto years before a tiny company in New Mexico, the now-legendary MITS, produced the first affordable, do-it-yourself computer—the *Altair*.

In March, 1977, Alan Kay and Adele Goldberg condensed a PARC technical report into an article, the title of which described both the dream and the reality of the Smalltalk branch of the PARC project: "Personal Dynamic Media" was published in a magazine named *Computer*, during a time when computer magazines were for specialists. Like Bush, Licklider, Taylor, and Engelbart before them, Kay and Goldberg did not talk of circuits or programs, but of media, knowledge, and human creative thought: [5]

> For most of recorded history, the interactions of humans with their media have been primarily nonconversational and passive in the sense that marks on paper, paint on walls, even "motion" pictures and television, do not change in response to the viewer's wishes. A mathematical formulation—which may symbolize the essence of an entire universe—once put down on paper, remains static and requires the reader to expand its possibilities.
>
> Every message is, in one sense or another, a *simulation* of some idea. It may be representational or abstract. The essence of a medium is very much dependent on the way messages are embedded, changed, and viewed. Although digital computers were originally designed to do arithmetic computation, the ability to simulate the details of any descriptive model means that the computer, viewed as a medium in itself, can be *all other media* if the embedding and viewing methods are sufficiently well provided. Moreover, this new "metamedium" is *active*—it can respond to queries and experiments—so that the messages may involve the learner in a two-way conversation. This property has never been available before except through the medium of an individual teacher. We think the implications are vast and compelling.
>
> *A dynamic medium for creative thought: the Dynabook*. Imagine having your own self-contained knowledge manipulator in a portable package the size and shape of an ordinary notebook. Suppose it had enough power to outrace your senses of sight and hearing, enough capacity to store for later retrieval thousands of page-equivalents of reference materials, poems, letters, recipes, records, drawings, animations, musical scores, waveforms, dynamic simulations, and anything else you would like to remember and change.

The Learning Research Group introduced students from the nearby Jordan Middle School in Palo Alto to what they called "interim Dynabooks." Nearly a decade before keyboards and display screens became familiar appliances, these children were introduced to a device no child and only a few computer scientists had seen before—an Alto computer set up to run Smalltalk. By using the mouse and the graphics capabilities provided by the hardware and software, these students were able to use Smalltalk to command the computer in much the same way that Papert's students in Cambridge, years before, had learned to program in LOGO by "teaching the turtle new words."

The screen was either a "very crisp high-resolution black-and-white CRT or a lower-resolution high-quality color display." High-fidelity speakers and sound synthesizers, five-key keyboards like Engelbart's, and piano-like keyboards were also available. The system could store the equivalent of 1500 pages of text and graphics, and the processor was capable of creating, editing, storing, and retrieving documents that consisted of words, graphic images, sounds, numbers, or combinations of all four symbol forms.

The mouse could be used to draw as well as to point, and an "iconic editor" (another Smalltalk innovation) used symbols that children who were too young to read could use to edit graphics; e.g., instead of typing in a command to invoke a graphics cursor, a child could point at a paintbrush icon.

The interim Dynabook could be used to read or write an old-fashioned book, complete with illustrations, but it could also do much more: "It need not be treated as a simulated paper book since this is a new medium with new properties. A dynamic search may be made for a particular context. The non-sequential nature of the file medium and the use of dynamic manipulation allows a story to have many accessible points of view; Durrell's *Alexandria Quartet*, for instance, could be one book in which the reader may pursue many paths through the narrative," wrote Kay and Goldberg.

The dynamic nature of the medium was made clear to users as they became acquainted with the toolkit for drawing, editing, viewing, and communicating. Smalltalk was not just a language, and the Alto system was not just a one-person computer. Together, the hardware, the software, and the tools for users to learn the software, constituted an *environment*— a small symbolic spaceship that the first-time user learned to control and steer through a personal universe.

The ability of the users to personalize their representation and use of information became clear as the children from Jordan Middle School experimented with changing the typefonts for displaying letterforms, and with changing the bit-maps of the computer to create and animate cartoon images in mosaics, lines, and halftones. The users not only had the capability to create and edit in a new way, but once they learned how to use the new medium they gained the ability to make their own choices about *how to view* the universe of information at their fingertips.

The editing capabilities of the Dynabook made it possible to display and change every object or description in the Smalltalk microworld. Text and graphics could be manipulated by pointing at icons and lists of choices—"menus" in software jargon—and multiple "windows" on the display screen made it possible to view a document or group of documents in several different ways at the same time. The filing capabilities made it possible to store and retrieve dynamic documents that could consist of any collection of objects that could be displayed and have something to do with each other. Drawing tools and painting programs made it possible to input information freehand as well as through the keyboard.

The structure of the Smalltalk language, the tools used by the first-time user to learn how to get around in the Dynabook, and the visual or auditory displays were deliberately designed to be mutable and movable in the same way: "Animation, music, and programming," wrote Kay and Goldberg, "can be thought of as different *sensory views* of dynamic processes. The structural similarities among them are apparent in Small-talk, which provides a common framework for expressing those ideas." A "musical score capture system" called OPUS and a graphic animation tool named SHAZAM were part of the Smalltalk-Dynabook toolkit.

In 1977, *Scientific American*'s annual theme edition was dedicated to the subject of "Microelectronics." Alan Kay's contribution to the issue, "Microelectronics and the Personal Computer," was the only article that directly talked about the meaning of this new technology for *people*. The magazine's editors summed up the piece in a two-sentence subtitle: "Rates of progress in microelectronics suggest that in about a decade many people will possess a notebook-size computer with the capacity of a large computer of today. What might such a system do for them?"

One of the first things Kay pointed out was the connection between the use of interactive graphic tools and the exercise of a new cognitive skill—a skill at selecting new ways to view the world. The metamedium which Kay still saw to be a decade or so in the future would only achieve its full power when people use it enough to see what it is about. The power that the 1977 prototypes granted to the human who used such devices was the power to create many new points of view.

This freedom to change one's view of a microworld, Kay believed, was one of the most important powers of the new kinds of representational tools that were being designed and tested in the late 1970s. In describing the way children learned to use the Smalltalk system, Kay also described something of the nature of the experience: [6]

> Initially the children interact with our computer by "painting" pictures and drawing straight lines on the display screen with the pencillike pointer. The children then discover that programs can create structures more complex than any they can create by hand. They learn that a picture has several representations, of which only the most obvious—the image—appears on the screen. The most important representation is the editable symbolic model of the picture stored in the memory of the computer. . . .
>
> One of the best ways to teach nonexperts to communicate with computers is to have them explore the levels of abstraction at which images can be manipulated.

Kay noted that when he gave the same tool that the children used as both an amusement and an entrance into Smalltalk programming to an adult artist, the artist started out creating various designs similar to those he was accustomed to making on paper. Eventually the artist discovered that the properties of the new medium, and his increasing facility for commanding those properties, made it possible for him to explore graphic universes that he could never have created with his old media: "From the use of the computer for the impoverished simulation of an already existing medium," Kay wrote, "he had progressed to the discovery of the computer's unique properties for human expression."

This freedom of viewpoint was only meant to be explored and demonstrated in a preliminary way in Smalltalk. It was Kay's hope that many new metaphors and languages would evolve as time went on, into what he called "observer languages": [7]

> In an observer language, activities are replaced by "viewpoints" that become attached to one another to form concepts. For example, a dog can be viewed abstractly (as an animal), analytically (as being composed of organs, cells and molecules), pragmatically (as a vehicle by a child), allegorically (as a human being in a fairy tale) and contextually (as a bone's way to fertilize a lawn). Observer languages are just now being formulated. They and their successors will be the communication vehicles of the 1980's.

Kay set forth his theories about personal computers as the components of a new medium for human expression, and compared the recent and future emergence of personal computers with the slower development cycles of past media. He also predicted that the changes in the human social order that were likely to accompany a new computerized literacy would be much more sweeping than the effects of previous media revolutions. The creation of a literate population would be the first reason for such a change. Out of that literate population, perhaps a few creative individuals would show the rest of us what could be achieved. He declined to predict the specific shape of these social changes, noting the failure of previous attempts at such forecasting: [8]

> We may expect that the changes resulting from computer literacy will be as far-reaching as those that came from literacy in reading and writing, but for most people the changes will be subtle and not necessarily in the direction of their idealized expectations. For example, we should not predict or expect that the personal computer will foster a new revolution in education just because it could. Every new communication medium of this century—the telephone, the motion picture, radio and television—has elicited similar predictions that did not come to pass. Millions of uneducated people in the world have ready access to the accumulated culture of the centuries in public libraries, but they do not avail themselves of it. Once an individual or a society decides that education is essential, however, the book, and now the personal computer, can be among the society's main vehicles for the transmission of knowledge.

The difference between a Dynabook of the future and all the libraries of the past, however, would depend upon the dynamic nature of this medium. A library is a passive repository of cultural treasures. You have to go in and dig out your own meanings. A Dynabook would combine the addictive allure of a good video game with the cultural resources of a library and a museum, with the expressive power of an animated fingerpaint set and a synthesized orchestra. Most importantly it would actively find the knowledge appropriate for the task of the moment, communicated in the form and language best suited to each individual who used it.

The intelligence of such devices—the reason that software breakthroughs in artificial intelligence research would someday have to intersect with the evolution of personal computers—would influence their ability to bring resources to the person who needs them. When the machines grow smart enough to communicate with eight-year-olds, then the question will shift from how to build a computer that people can easily use to what we all do with that kind of power.

What if libraries were able to find out what most interests you and what you most need to know, and showed you how to find what you wanted? What if you could say to the library: "I wonder what it would be like to live in Baghdad of the Caliphate?" or "I wonder how it feels to be a whale?" and expect the library to *show* you? Do you like Van Gogh? How about a simulation of the fields outside his house? Would you care to sit in with Louis Armstrong or Wolfgang Mozart? What would it do to the world if we could all see how everybody else lived and share in their cultures?

If the first effect of the coming metamedium was likely to be the creation of a literate population who shared a new freedom to use symbols and to choose how to view information, then the second effect lay in the power that would be unique to this medium—the power of *simulation*. Simulation is *the power to see what you imagine*, to create worlds that obey your command. The computer can build instant sensory representations. The user/programmer explores a universe that *reacts*, in which the degree of the user's power depends upon and grows with one's understanding of the way the worlds work.

The power of simulation to empower the imagination and give form to whatever can be clearly discerned in the mind's eye is what makes this kind of device a "fantasy amplifier." Although there are several homilies that are entitled to be called "Kay's First Law," the statement that he most often calls "Kay's Second Law" is: "Any time you build a fantasy amplifier, you have a winner." His reasoning is that game playing and fantasizing are metaphors for the kind of skill people need to get around in the world.

"We live in a hallucination of our own devising," Kay is fond of saying. But our illusion is so complex, so much of the world we experience appears to be beyond our control, and the operating manual is so difficult to find, that we all tend to get locked into the way our families, societies, and cultures see the world. "We can't exist without fantasy," Kay asserts, "because it is part of being a human. A fantasy is a simpler, more controllable world."

And by practicing how we would control a simpler version of the world, we often figure out how to operate the world outside the fantasy. A game is both controllable and challenging. It is entered vicariously, purposefully, and with an open mind about the outcome. Sports and science and art all involve vicarious, purposeful fantasies in that sense. That's why he feels that video games were not a fad but a precursor to something with much more profound power. And that is the most likely reason why he joined Atari Corporation.

The power of simulation is not necessarily or exclusively a beneficial one, as the legends of today's system-crashers, obsessed programmers, and dark-side hackers attest, and as Kay warned in his *Scientific American* paper: [9]

> The social impact of simulation—the central property of computing—must also be considered. First, as with language, the computer user has a strong motivation to emphasize the similarity between simulation and experience and to ignore the great distances that symbols interpose between models and the real world. Feelings of power and a narcissistic fascination with the image of oneself reflected back from the machine are common. Additional tendencies are to employ the computer trivially (simulating what paper, paints and file cabinets can do), as a crutch (using the computer to remember things that we can perfectly well remember ourselves) or as an excuse (blaming the computer for human failings). More serious is the human propensity to place faith in and assign higher powers to an agency that is not completely understood. The fact that many organizations actually base their decisions on—worse, take their decisions from—computer models is profoundly disturbing given the current state of the computer art. . . .

The *fact* of simulation is so seductive to human perception, and so potentially useful in "real-world" applications, that its widespread use is inevitable, once personal computers grow sophisticated and inexpensive enough. The ethics of *how* and *for what purposes* simulations should and should not be used are only beginning to be formulated. The historical events, debates in PTAs and legislatures, and growth in public concern that will accompany the introduction of this medium will help determine the shape of the future ethics of simulation. The best place to look for expert guidance, Kay suggests, might be to those of us who are the least prejudiced by precomputer ways of thinking: [10]

> Children who have not yet lost much of their sense of wonder and fun have helped us to find an ethic about computing: Do not automate the work you are engaged in, only the materials. If you like to draw, do not automate drawing; rather, program your personal computer to give you a new set of paints. If you like to play music, do not build a "player piano"; instead program yourself a new kind of instrument.

The way we think about computers—as machines, as systems that mimic human capabilities, as tools, as toys, as competitors, or as partners—will play a large part in determining their future role in society.

In the conclusion of his article, Kay cautions against the presumptions of present-day minds about what the minds of future generations may or may not choose to do with the instruments past generations have worked to create: [11]

> A popular misconception about computers is that they are logical. Forthright is a better term. Since computers can contain arbitrary descriptions, any conceivable collection of rules, consistent or not, can be carried out. Moreover, computers' use of symbols, like the use of symbols in language and mathematics, is sufficiently disconnected from the real world to enable them to create splendid nonsense. Although the hardware of the computer is subject to natural laws (electrons can move through the circuits only in certain physically defined ways), the range of simulations the computer can perform is bounded only by the limits of human imagination. In a computer, spacecraft can be made to travel faster than the speed of light, time to travel in reverse.
>
> It may seem almost sinful to discuss the simulation of nonsense, but only if we want to believe that what we know is correct and complete. History has not been kind to those who subscribe to this view. It is just this realm of apparent nonsense that must be kept open for the developing minds of the future. Although the personal computer can be guided in any direction we choose, the real sin would be to make it act like a machine!

Because he started out young in a field that was young itself, Kay was one of the first of the generation of infonauts, the ones who grew up with the tools created by the pioneers, and who have been using them to create a medium for the rest of us. One of the things he learned at ARPA and Utah, SAIL and PARC, Atari and Apple, was that putting together a group of talents and leaving them alone might be the most important ingredient in invoking the breakthroughs he'll need to complete his dream.

People are beginning to wonder what Kay, now at Apple, intends to do next. "I would imagine he feels more than a little frustrated," said Bob Taylor, in 1984, referring to the fact that Alan Kay hadn't produced anything as tangible as Smalltalk in a number of years. A hotshot programmer at Apple put it differently: "He deserves to be called a visionary, because he is. And I love to hang around him because he knows so much about so many things. But it gets a little tiring the third time you hear hear him say, 'We already did that back in '74.' "

Atari was the first institution where Alan Kay played a significant

role but didn't make any breakthroughs. Because of what happened—
or didn't happen—with the Atari team, he probably learned that being
a member of a team, albeit an inspirational, even visionary member,
doesn't necessarily mean that he is cut out to be a good leader. Before
we explore the end of the dream at Atari, however, another infonaut
by the name of Brenda will give us a glimpse at part of what Kay and
his cohorts attempted to accomplish.

CHAPTER TWELVE

Brenda and the Future Squad

To those of us who don't live and work in futurist sanctums like ARC, PARC, Atari, or Apple, such activities as flying through information space or having first-person interactions with a computer are hard to imagine in terms of what one would like to do on a Friday night. There simply aren't any analogous images available in our cultural metaphor-bank. Is it like watching television? Playing a video game? Searching through an infinite encyclopedia? Acting in a play? Browsing through a book? Fooling with fingerpaints? Flying a plane? Swimming?

My initial encounter with Alan Kay led me to several of the people who worked for him at the time, and I eventually ended up spending more time with Brenda Laurel and colleagues than I did with Alan. Brenda and her friends were interested in the same questions that puzzled me: What would it *feel like* to operate tomorrow's mind-augmenting information-vehicles? My first experience with their work took place in a guarded, well-equipped room in Sunnyvale, California, home of Atari Systems Research Group. The following brief scenario is taken from my notes of that first observation:

The world was gray and silent before Brenda spoke.

"Give me an April morning on a meadow," she said, and the gray was replaced by morning sunshine. Patches of cerulean sky were visible between the redwood branches. Birds chirped. Brooks babbled.

"Uhhmm . . . scratch the redwood forest," Brenda continued: "Put the meadow atop a cliff overlooking a small emerald bay. Greener. White-caps."

Brenda was reclining in the middle of the media room. "The background sounds nice," she added: "Where did you get it?"

"The birds are indigenous to the northern California coast," replied a well-modulated but disembodied female voice: "The babbling brook is from the acoustic library. It's digitally identical to a rill in Scotland."

"There's a wooded island in the bay," continued Brenda, looking down upon the island that instantly appeared below her where only green water had been a moment before. She surveyed the new island from her meadow atop the cliff above the bay, then spoke again: "Monterey pine, a small hill, a white beach. Zoom into the beach. Let's walk up that path. There's a well under that banyan tree. I want to dive in and emerge bone-dry in the Library of Alexandria, the day before it burned."

A few groups on the leading edge of cognitive technology have been trying to find images to help them in their effort to materialize a mass-marketable version of Bush's Memex, Engelbart's Augmentation Workshop, and Kay's Dynabook. Those people who are attempting to design these devices share an assumption that such machines will evolve from today's computer technology into something that will probably not resemble the computers we see today. Ideally, we won't *see* these hypothetical evolved computers of tomorrow, because they will be invisible, built into the environment itself.

Try to imagine a computer that is nowhere to be seen, and is set up to attend to your every wish, informationally speaking. You enter a room (or put a helmet over your head), and the room (or the helmet) provides multisensory representations of anything, real or imaginary, you can think of asking it to represent. Science fiction writers of the past decades have done their share of speculating on what one might do in such a representationally capable environment. You could, for example, go skiing in the Alps with wraparound full-color, three-dimensional visual display, authentic panaphonic soundtrack, biting cold air, ultraviolet-rich high-altitude sunshine, spray of powder snow on your cheeks, the feeling of skis beneath your feet, of being impelled down a slope.

But you shouldn't have to limit your use of such a universal information medium to a real terrestrial experience. You could explore a black hole in a neighboring galaxy, navigate through your nervous system, become a Connecticut Yankee in King Arthur's court. If you want to extend your senses into the real world in real time, you can look at quasars with x-ray radiotelescope vision, CAT scan everything you see, hover above the earth in a weather satellite, zoom down to take an electron microscopic look at the microbes on a dust mote on a license plate in Kenya.

If you want to communicate with one person or an entire on-line network, you have all the media at your disposal, along with additional "dialogue support tools" to augment the interaction. Or the interaction might be private, limited to you and the informationscape—for reasons of work or play.

Perhaps you want to know something about blue whales. Everything written in every magazine, library, or research data base is available to you, and an invisible librarian is there to help you, if you wish: Just focus your eyes on a reference and it fills the screen. Ask the librarian questions about what you want to know, or allow it to ask you questions. But you don't have to just read about whales. You can listen to them, watch them, visit them. Just ask, and you'll be underwater, swimming among them, or in a helicopter, watching them while you hover above crystalline Baja waters.

The experience won't be strictly passive. You can act out the role of a whale or Louis XIV (or Genghis Khan, if that is your taste) in a simulated video encounter and make decisions about the outcome of that encounter. Paint palettes, text editors, music and sound synthesizers, automatic programming programs, and animation tools will give you the power to create your own blue whale or ancient Mongolian micro-worlds and romp around in them.

Since MIT, Lucasfilm, and Evans & Sutherland were in the bidding for Kay's services when he left Xerox, one can safely assume that Atari must have offered him something more. Although his obvious desire was to run an advanced software shop, Kay knew that his next software dream would require very advanced hardware. "You want hardware de-signers? We'll get you hardware designers," you can imagine them saying. Atari got him nothing but the best—including Ted Hoff, the legendary Intel scientist who was the leader of the team who invented the micropro-cessor chip. Kay assembled his own software research team.

Brenda Laurel joined Atari Systems Research Group after a stint in their educational marketing division. When I first met her, she was involved in a research project that she insisted defied verbal description. She invited me to watch a a special kind of brainstorming session they were just beginning to explore.

The Atari research building was in a typical Sunnyvale flatland industrial park, with the usual high-tech high-security trappings—twenty-four-hour guards, laminated color-coded nametags, uniformed escorts. It was here that I joined Brenda and several of her colleagues in a group-imagination exercise connected with what they called a media-room project.

Brenda signed me in, walked me through the gray-walled, gray-carpeted corridors, and brought me to a large room, bare except for a few industrial-modern couches and chairs, a videotape setup, and two whiteboards. Inside the room were Eric Hulteen, the project leader; Susan, a red-haired, soft-spoken young woman; Scott, a quiet, spaced-out preppie type; Don and Ron Dixon, the robotic experts; Craig, a somewhat skeptical, bearded hacker; Jeff, Tom, Brenda, and Rachel, who was videotaping the event.

Rachel was short, had a crewcut, wore a tank-top tee shirt, purple blousy harem pants, and no shoes. Don and Ron were twins. A few in the group could have been as young as twenty-three or twenty-four, the oldest was no older than thirty-five. Jeans and sandals were the dominant costume. Nobody wore a tie. Nobody had acne or speech impediments. Nobody wore a plastic penholder.

As it was explained to me by Brenda and by Eric, whose project it is, a media room is an information terminal that a person can walk around inside—a place where you can communicate directly with the machine without explicit input devices like keyboards. The room itself is set up to monitor human communication output. This presumes that all of the hardware and software that are now in experimental or developmental stages will be working together to do what a good media room does—without bothering the person who uses it with details of its operations.

Eric came to Atari from MIT's Architecture Machine Group, an innovative group led by Alan Kay's old friend and Atari consultant Nick Negroponte. The idea of "spatial data management" that came from the MIT group was a response to the problem of finding a way to navigate the huge new informational realms opened by computers, by adopting the metaphor of *information space* that the user can more or less "fly" through. The dominant metaphor in software design viewed large collections of information through the lens of the well-known "file-cabinet" metaphor, in which each piece of information is regarded as part of a "file folder" that the user locates through traditional filing methods. But what if the collection of information could be displayed visually and arranged spatially, so the user could have the illusion of "navigating" through it?

Perhaps the most well-known demonstration of this metaphor was the "Aspen Map" created by Negroponte's group. To use this map, you sit in front of a video screen and touch the screen to steer your way down a photographic representation of the streets and houses of Aspen, Colorado.

A computer-directed videodisc system connects the video steering controls to a very large collection of photographs of Aspen. The computer translates your position and your commands into the correct sequence of photographs. If you decide to look to the left, the screen shows the streets and houses that are located to the left of this position in the real city. If you decide to stop and take a closer look at one of the houses that are specially marked, or even open the door and look inside, you can do so.

The kind of simple branching structure of a city's streets represents only the most basic kind of information base that can be represented spatially. The most important aspect of this idea doesn't have to do with road maps—although this is obviously a good way to learn how to get around in a town you've never seen before. The important point is that some information domains can be organized around a spatial metaphor, creating a coherent environment that each user can move around in by following his own particular path. A reference work for someone trying to find the problem in an automobile engine or the plumbing system of a nuclear submarine could just as easily be mapped in such a way.

Whether they came from MIT, Carnegie-Mellon, or another video game manufacturer, every person in Kay's Atari group represented the cream of the crop of the best young minds in fields ranging from robotics to holography to videodisc technology to artificial intelligence to cognitive psychology to software design. The necessary hardware components of the media room will become available, everyone hopes, by the time the really tricky part—the software design, construction, and debugging—is on its way to completion.

The person inside a full-scale media room will have 360-degree visual displays of some sort—high-resolution video or holographic images—computer-generated and archived. Images can be retrieved from a library (and added to the library), or they can be constructed by the person or by the computer. There will be a total-sound audio system ranging into ultralow and ultrahigh frequencies. But the most important element is not in the sensory displays, which involve straightforward if now-expensive technology, but in the software—in the way the room is designed to "know" what to do.

If the media room is to be the universal medium, the room itself must be able to see and hear the person inside, and "understand" what it sees and hears well enough to carry out the person's commands. Ideally, it should understand the person it is dealing with well enough to actively *guide* the fantasy or the information search, based on its knowledge of personal preferences and past performance. Bioelectronic sensors built into the floors will keep track of the user's mood. The only thing the room is *not* presumed to do is to read minds.

One of the ways to describe a media room is "a computer with no interface," or "a computer that is *all* interface." When the computer interface disappears, you are not at the control panel of a machine, but walking over the Arctic ice, or flying to Harlem, or looking through a book in a musty old room. How does one envision the capabilities of a technology that doesn't exist yet? How do you deal with an invisible computer? If you don't have to worry about how to tell it what to do, and if its information-representation capabilities are too large to worry about, the question shifts from the tool to the task: "Okay, now that I can go anywhere, including places that don't exist, where do I want to go?" Brenda, Eric, and their colleagues wanted to know what new communication styles people might adopt in response to such a system. Most of all, they wanted to know how it would *feel* to use such a system.

The night I watched her and her colleagues fantasize in that room in Sunnyvale, Brenda's idea was to plan the uses of a future technology of this sort by using the same kinds of tricks that actors use to create imaginary spaces: "Magical kinds of things can happen through improvisation," she told the group, "because it can trick you into revealing preverbal ideas. What we each bring to this is our capacity to have inspirations in real time."

The first improvisations were warm-up exercises. Brenda's trip to the Library of Alexandria was followed by Scott's visit to a hypermirror that showed him what he looked like in the infrared and gave him a real-time scan of his brain metabolism in sixteen colors. He watched the colors of his thought processes as he watched the colors of his thought processes.

Then the group decided to make Eric play the role of the person using the system, while everyone else improvised roles as the components of the media room—input to the user's vision, mobility, hearing, emotions, thoughts. In the first try, everyone got into their role with such enthusiasm that Eric was literally swarming with people mimicking him, whispering advice, grimacing. He spent his time rather defensively trying to figure who did what. It was like a combination of twenty questions and charades, but it revealed something about the bewilderment of even a technically sophisticated computer user when faced with a system that does not explain itself but simply *acts*.

In the next experiment, Susan, acting as the person in the middle of such a system, decided to try to take control of the elements, and discovered that all the roles of the different components could be changed radically by adding a "help agent." The help agent oriented the user by saying things like "ask her [the librarian component] about a place," or "ask him [the user-memory component]—he knows what to look for." The idea was to create a kind of "informational butler" that would observe both the user and the information system, keep a record of that individual's preferences, strengths, and weaknesses, and actively intervene to help the user find or do what the user wanted to find or do.

The next day, several of the crew were going to Southern California, to see what a prominent university cognitive science department could tell them about designing machines that people can use. About a week later, Brenda and I talked about what she had learned from the cognitive scientists, and the improvisation exercise.

"The cognitive science people are looking at human-machine interactions. Naturally, the hired hackers got into the act when the subject of the discussion was how to teach secretaries to use a file management system. One of the programmers at the staff meeting summarized the problem by asking, 'How do we get a secretary to understand that slash–single quote–DEL will delete a file?' That was his understanding of the human interface—a matter of figuring out how to adapt a human to the esoteric communication protocol some programmer built into a machine."

That part of a computer game that makes the user step outside the game world, that doesn't help the user to participate in the pleasure of the game, but acts as a tool for talking to the program—that's where distance comes in. That's what happens to the secretary when the programmer tells her that slash–single quote–DEL means "erase this." She doesn't want to ask the computer to erase it; she simply wants to erase it.

What Brenda was getting at seemed so strange and ran so counter to everything I had been taught that it took a while for it to sink in: In essence, she was saying that when it comes to computer software, the human habit of looking at artifacts as *tools* can get in the way. Good tools ought to disappear from one's consciousness. You don't try to persuade a hammer to pound a nail—*you* pound the nail, with the help of a hammer. But computer software, as presently constituted, forces us to learn arcane languages so we can talk to our tools instead of getting on with the task.

"The tool metaphor gets in the way when it is applied at the level of the larger system that includes the human operator," Brenda explained. Even though your programmer gives you a file management system that is functional in a tool-like way, the weird way the human is forced to act in order to use the tool creates an unnecessary distance between the action that the human is required to perform and the tool's function.

"We also know, however, that there is *another* set of computer capabilities that aren't at all tool-like. Games and creating art, for example. So what is it that a computer does, in that case? My answer is that its function is to *represent* things. Which, in the case of art or games, means that the function is the same as the outcome, because in art or games, representation is at least part of the outcome."

Kids don't play video games by the hour because it is a good way to practice hand-eye coordination, or for any other reason besides the sheer pleasure of playing. On the other hand, nobody uses a word processing program out of sheer enjoyment of using the program; they use a word processor because they want to write something. In the case of the game, the process is most important. In the case of the word processor, the outcome is most important. In the video game, there is no separation from the user/player and the world represented in the game. In the word processor, the command language of the software creates a distance between the user and the task.

"One strategy in our research is to find out how to eliminate the part that keeps us distanced," Brenda explained. "I want to reach my hands right through the screen and do what I want to do," she added, with the kind of passionate conviction I hadn't encountered since Engelbart got that faraway look in his eyes and started talking about what humankind could do with a true augmentation system. "I don't want to enter a bunch of commands," Brenda insisted. "I might not even want to *speak* a bunch of commands, if I have to speak them in a way that is different from the way I normally talk. I want *first-person* interaction. Great. But first I have to do away with all this stuff between me and the outcome.

"What metaphors *haven't* been used? Maybe the interface is a barrier. I think that it is more than a technological question. You can't expect to solve a problem by building better interfaces if the whole idea of interface is based on an incomplete metaphor. To use a real artsy metaphor that will probably break down under scrutiny, I like to look at the computer as a system for making magic portals. Like that moment in *The Wizard of Oz* when Dorothy opens the door and everything changes from black and white to color. *That* is what I want to happen—perceptually, cognitively, emotionally. The portal is an interim metaphor to me. We need something richer. I'm looking for something that will click into place and reexplain the idea of the interface.

"I want to make a fantasy that I can walk through," Brenda explained. "That is what an adventure game tries to do. Long before computers were available to regular folks, hackers on large mainframe computers were hooked on adventure games. Now there are adventure games that you can play on your home computer. What happens when you try to build a *first-person* adventure game?

"The first thing I do in this game I want to walk around in is to *look* at it. Maybe there are some graphics on the screen. Perhaps the screen is all around me. Maybe there is some text to read, or a sound track that reads it to me. All of these are important technical aspects, but they are peripheral to my concern. All the screen and speakers do is to establish an environment. Once I look around the environment, however, I want to interact with it.

"Let's say that the environment of this fantasy is something that a science fiction writer of the first caliber invented. Say it's a planet that I'm exploring for the United Federation of Planets. I start walking through this world. Today, with the state of the interface art as it is, if I want to move to the north and turn over a stone, I'd tell the computer, 'Move north. Turn stone.' Note that I have to *tell the computer*. I've just stepped out of the fantasy. And you destroy a fantasy when you step out of it.

"What kind of system enables me to simply move north and pick up the damn stone? I don't think it's just a question of making the environment lifelike. It isn't just a technical question for a fancier projector to solve. It's a question of how the world is established when it is constructed. How the author establishes the way in which people can relate to it.

"Maybe I can look around the planet until I find a guide. Remember the 'help agent' in the media-room improvisation? This description of walking around a world sounds a lot like a theatrical improvisation. You walk up to the stage, and the director says, 'Okay, this is a new planet. You play an explorer. Go.' Nine times out of ten, something like that dwindles away, but if you are lucky you discover something useful about the character. Very rarely do you look back and say, 'That was a wonderful story.' "

According to Brenda's theory, the reason the story is rarely memorable, even in a good improvisation, is because the actors are forced to use part of their mind to think about being playwrights. To achieve an excellent dramatic outcome the actor has to think about his character and manipulate the plot line at the same time, so that it all comes out in an interesting way. Unless you're a genius, you have to trade part of your acting power in order to think about the play. And you can't do a great job crafting a drama if you have the acting ball to juggle.

"This is where I think a computer can assist us," Brenda insists: "I think one answer is to put the smarts of a playwright into a first-person fantasy-creating system.

"It has to be built into the way the imaginary world is constructed. Sitting on top of all your graphics and voice recognition and speech synthesis is an expert system that can make informed decisions about the potential of dramatic situations, using a large enough base of knowledge about the possible situations that can arise and a set of rules for sifting through that knowledge base."

Less fantastic, but nonetheless powerful versions of the "expert system" Brenda was talking about do exist now—and in the next chapter we'll take a look at what another infonaut thinks about the potential of these "knowledge-transferring" programs. The hypothetical variation Brenda was describing would be able to learn from experience—experience with the individual who is using it or with everybody who has ever used it. Brenda thinks that such a program could approach the kind of analysis that a drama critic does. "Maybe we can put Aristotle's rules for good drama in the system to start."

Right now, there are expert systems in existence that can help doctors to diagnose diseases. Those systems are able to apply diagnostic rules adapted from human doctors to a large collection of data, a knowledge base, regarding known symptoms. Substitute drama for disease, and the elements of drama (like universality and causality) for symptoms, and the automatic drama expert in our fantasy will be able to pick out the most dramatic responses and consequences for actions that the player performs, and weave them back into the fantasy. It's an idea that seems to be as far ahead of today's entertainment software as Alan Kay's Dynabook was ahead of the computer hardware of the 1960s.

Assume that you can simulate a medieval castle and give an audience member a 360-degree, first-person interactive presence in the dramatic action, so that every time you step into the *Hamlet* world as Horatio or Hamlet or Ophelia, you make different choices that affect the outcome.

Artificial intelligence research tells us that you don't have to specifically store all the possible events that could occur in a giant data base if you can structure the representation of the world in such a way that its characteristics are formulated as tendencies to go in certain directions. When you pick up a stone, for example, you are likely to find crawly things under it.

Leaving aside technical arguments about the feasibility of constructing such a system, Brenda is most concerned about what effects the experience of encountering a system like the one she described might have upon our emotions as well as our cognitions: "How does it feel to experience a world like that? How does it change my perception to walk through its portals? How do I find out where the edges are? What kinds of transactions can I have with this world?"

The experience Brenda described is the experience at the human interface—where the mind and machine meet. The interface hardware and software are what computer people call the "front end" of the system. The back end is what the system needs in the way of smarts so that outcomes end up being dramatically pleasurable. Right now, you can wander around in an adventure game and gather treasure and kill monsters and finish by winning or being killed. There isn't a sense of unfolding drama. In order for the front end of an adventure game to convey that sense of direct, first-person drama, it would have to be based on a very sophisticated back end.

"You use existing technology to make the scenes branch according to your decisions, but that doesn't converge on a dramatic outcome, except in the most mechanical way. But you could take the same world with the same characters and the same elements and add this sense of drama, and come out with something that would be more like experiencing a drama at first hand.

"The kind of system I'm describing has to be able to find out what I want by remembering what kinds of things I have paid attention to. The system has to have a good enough model of me, and memory of how I have acted in the past, to make good guesses about how I'm likely to act in the future.

"I've tried to describe an element from the simplest thing that I think my colleagues and I will actually be able to do in the near future. Let's look down the road ten years. Say we really get the system working and we know how to synthesize dramatic outcomes and orchestrate sound tracks and images and give the person who uses the system a way to affect the representations.

"We can think of such a system not only as a medium for an interactive fantasy but as a kind of an interface to information that is not fantasy. What if the world, instead of planet X or Shakespeare's Denmark, is the world of whales or the world of chemical reactions? That's a powerful idea that we can see at work right now in the best of contemporary educational software."

She offered the example of a game in which the players experience the fantasy of being cadets on a starship. Each cadet would be responsible for running part of the ship. The players can choose whether they want to specialize in navigation or propulsion or life support or computer systems. In real time, they run their parts of the ship. Then something goes wrong—the life-support systems are threatened, the reactor is malfunctioning. Or something interesting occurs—the exobiologists have spotted a planet to investigate. The players have to find out what to do and how to do it. In the first person.

"Now let's look at it from the point of view of drama theory," she proposed. "You accept very easily the idea that I am a space cadet. I accept it too. This is what happens when a master actor impersonates a character. When I am impersonating someone, *all* of me is impersonating that character. What has to go away, to disappear from my own behavior to make that possible? The idea that I am me—the person who doesn't know what I haven't learned—has to go away. The same idea that often gets in the way of learning anything new.

"A willing suspension of disbelief that accompanies a first-person simulation enables the person who participates to feel what it would be like to have greater personal power. A world like that shows us what it's like not to have the limitations that we think we have in everyday life. When we see how much a kid learns about predicting trajectories and the rules of bodies in motion from playing even a simple video game, I think it is easy to see educational potential in using these 'fiction environments' as the door to worlds of information that are as useful or healthy to know as they are fun to learn about."

Of course, by this time, I was asking the same question that most of the people reading this chapter must be asking: "When are we going to play with these 'fiction environments'? How close is Atari to releasing actual products based on this research?"

The answer, unfortunately, is that it is unlikely that Atari is ever going to translate this research into consumer products. Six months after I talked to Alan Kay and observed Brenda Laurel's research group, the Systems Research Group was fired en masse. Brenda and Eric were given five minutes' notice. Alan resigned and went to Apple shortly thereafter. Once again, as in the case of ARC and PARC, it seemed that the management of the corporations that nurtured the most exciting research in interactive, mind-augmenting computer systems seemed to fail miserably when it came to the point of developing products.

After she was fired, Brenda was a lot more willing to talk about the pressures of doing long-term research for a consumer-product-oriented company. In her opinion, the explanation for the demise of Atari Research, and the dramatic reversal of Atari Corporation's fortunes that led to the drastic cutback, is a simple one. "The Warner people" (who owned Atari), she claims, "never knew anything about innovation. The people they hired to run Atari were from Burlington Industries, Philip Morris, Procter and Gamble—dog food boys. How often does dog food change?"

Before she was in Systems Research, Brenda was in marketing. She claims that she told Raymond Kassar (former CEO of Atari) that "what people are going to want from us is not more deadhead entertainment, but stuff that helps their minds grow. The largest market of all is the market for personal power, for new equivalents to opposable thumbs."

Augmentation visionaries like Engelbart, prophets of interactive computing like Licklider, and infonauts like Alan Kay and Brenda Laurel tend to talk in grand terms about the ultimate effects of what they are doing—the biggest change since the printing press or even since the opposable thumb. They all seem convinced that their projections will be vindicated by a technology that will inevitably come into existence despite the myopia of institutions like SRI, Xerox, and Atari.

With the increasing power of home computers, and the growing demand for entertainment and educational software, it seems likely that smaller groups, working in entrepreneurial organizations rather than academic or large-scale product-oriented institutions, will produce the fantasy amplifiers and mind augmentors of the near future. One of the most controversial areas of entrepreneurial research is in the field of applied artificial intelligence. The subject of the next chapter is involved in the commercial development of those intriguing programs that Brenda mentioned, the so-called expert systems that originated in the pure research that is being conducted at MIT and Stanford, and which seem to be invading the world of commercial software.

If intellectual augmentation and fantasy amplification are two of the possible paths that the human-computer relationship are likely to take in the next ten to twenty years, the most intriguing path that current research may lead to in the more distant future has to do with the possibility of using computers as tools for transferring knowledge. Avron Barr and his colleagues at TeKnowledge seem convinced that the commercial and humanitarian potential of software tools such as expert systems is much greater than most people suspect, or are yet willing to acknowledge.

CHAPTER THIRTEEN:

Knowledge Engineers and Epistemological Entrepreneurs

. . . it is extremely important that the development of intelligent machines be pursued, for the human mind not only is limited in its storage and processing capacity but it also has known bugs: It is easily misled, stubborn, and even blind to the truth, especially when pushed to its limits.

And, as is nature's way, everything gets pushed to the limit, including humans. We must find a way of organizing ourselves more effectively, of bringing together the energies of larger groups of people toward a common goal. Intelligent systems, built from computer and communications technology, will someday know more than any individual human about what is going on in complex enterprises involving millions of people, such as a multinational corporation or a city. And they will be able to explain each person's part of the task. We will build more productive factories this way, and maybe someday a more peaceful world. We must keep in mind . . . that the capabilities of intelligence as it exists in nature are not necessarily its natural limits.[1]

Are future computers going to become tools for extending the power of our minds, or are they going to evolve into a new kind of intelligent species that operates far beyond the limits of biological intelligence? Avron Barr, author of the statement quoted at the beginning of this chapter, is exploring one of the most potentially explosive areas of human-computer evolution—the field that has come to be known as "knowledge engineering."

To me, Barr's specialty seems to be rooted in the same idea that goes back to Licklider and Bush—the inevitability of a human-computer symbiosis. But to many other people, the idea of artificial intelligence seems to be fundamentally different from augmentation, in that the artificial intelligentsia appear to be more interested in *replacing* human intelligence than in extending it.

Knowledge engineering is but one part of that ever-expanding area of hardware and software research that constitutes the field of AI. Unlike other artificial intelligence researchers, Avron Barr is not concerned with systems that can direct an optical sensor to recognize visual patterns, or help a speech-recognition system to understand natural language, or direct a robot in the task of climbing stairs. He and his colleagues are trying to build systems that can transfer knowledge from experts to novices and that can use the transferred knowledge to help people make decisions about specific problems.

Barr's specialty seems to bridge the gap between those who see the future of computers in terms of "mind tools" and those who see it in terms of "the next step in the evolution of intelligence." Like the other people I met who have been involved in building tomorrow's software tools, Barr has a firm belief in the epochal quality of the changes we will face when these experiments filter down to the level of public experience. For example, consider the following scenario:

A general practitioner in a small town in the Southwest was awakened late one night by an emergency call—a six-year-old girl had been admitted to the local hospital. She was comatose, and she had a high fever. The doctor ordered all clinical tests that were available at that hour in a one-hospital town and called in the pathologist. The symptoms, and the results of the first tests, weren't like anything the GP or the pathologist had seen before. Drugs were available—the pharmacy was well equipped, even if specialized expertise was in short supply. But which drug?

Choosing the proper antibiotic from the hundreds of possibilities was a matter of life and death for the little girl, and neither the GP nor the pathologist was comfortable about staking the young patient's life on guesswork. They took their laboratory results over to the local community college, where one of the young programmers who always seemed to be around in the middle of the night used a microcomputer and a telephone to put them in contact with an expert in Palo Alto, California, who knew just the right questions to ask about a case like this.

"Has the patient recently had symptoms of persistent headache or other abnormal neurologic symptoms (dizziness, lethargy, etc.)?" asked the specialist in California.

"Yes," replied the local attending physician.

"Has the patient recently had objective evidence of abnormal neurologic signs (nuchal rigidity, coma, seizures, etc.) documented by physician observation or examination?"

"Yes," replied the pathologist.

With the help of clues provided over the telephone by the expert, the local doctors were able to administer one more test that narrowed their search for the disease-causing organism down to one of the three possibilities suggested by the specialist. There were drugs on hand for treating the infection that the long-distance expert had helped them pinpoint. The little girl recovered. The doctor, the pathologist, and the child's family were grateful.

The specialist, a computer program named MYCIN residing in a mainframe computer at the Stanford Medical Center, chalked up another diagnostic triumph to its already impressive record.

Although this particular story is fictional, the dialogue is an excerpt from a real MYCIN consultation. The program does indeed exist, and is in use as a strictly experimental diagnostic assistant. It is an example of a whole range of new computer programs known as *expert systems* that are now serving as intelligent assistants to human experts in fields as diverse as medicine and geology, mathematics and molecular biology, computer design and organic chemistry. Expert systems are just the first of a whole new variety of software probes that infonauts like Avron Barr are launching into the unknown regions of human-machine relationships.

These systems are both research tools and commercial products. A program called PROSPECTOR has recently helped pinpoint a molybdenum deposit worth tens of millions of dollars. A program named DENDRAL, which started out as an artificial intelligence experiment, is now owned by a consortium of chemical companies, whose chemists use it to design and synthesize potentially useful new compounds.

One important difference between an expert system and other kinds of computer programs is that the program does not simply provide answers to questions, the way a calculator provides the solutions to equations. Expert systems do, of course, suggest answers, and eventually they will venture answers accompanied by a numerical statement of "confidence" in the answer. But they do more than that. The most important part of an expert system is in the *interaction* between the program and the person who uses it.

The human who is faced with a specialized problem can consult the specialized program, which is able to ask the human questions of its own regarding the particulars of the problem. The consultation is a *dialogue* that is tailored to the specific case at hand. The program simulates the decision processes of human experts, and feeds back the results of that process to the human who consults it, thus serving as a reference and guide for the person who uses it.

Expert systems as they exist today are made of three parts—a base of task-specific knowledge, a set of rules for making decisions about that knowledge, and a means of answering people's questions about the reasons for the program's recommendations. The "expert" program does not know what it knows through the raw volume of facts fed to the computer's memory, but by virtue of a reasoning-like process of applying the rule system to the knowledge base; it chooses among alternatives, not through brute-force calculation, but by using some of the same rules of thumb that human experts use.

Statistics about how often experts turn out to be right are the ultimate criteria for evaluating expertise—whether the expert is a person who studied for years, or a computer program that was literally born yesterday. The methodology for conducting such an evaluation was suggested in the 1950s, by Alan Turing. The "Turing test" bypasses abstract arguments about artificial intelligence by asking people to determine whether or not the system they are communicating with via teletype is a machine or a person. If most people can't distinguish a computer from another human, strictly by the way the other party responds to questions, then the other party is deemed to be intelligent. A similar strategy has been employed to judge the efficacy of expert systems. Why not just ask some human experts to distinguish human from machine diagnoses?

One experiment conducted by the Stanford Medical School began by submitting to MYCIN the case histories of ten patients with different types of infectious meningitis. At the same time, eight human physicians, including five faculty specialists in infectious diseases, a research fellow, and a resident, were given the same information that had been fed to MYCIN. MYCIN's recommendations were sent, unidentified, along with the human physicians' recommendations, also unidentified as such, and a record of the therapy the patients actually received, to eight non-Stanford specialists. The outside specialists gave their highest rating to MYCIN.[2]

In the 1980s, there is little question that expert systems can be highly effective, if not superior to human expertise, in certain highly specialized fields. Twenty years ago, few people, even inside the artificial intelligence research community, were confident that it could be done at all. The normally "pure" research field of artificial intelligence strayed into this potentially controversial area of applied AI, as it was bound to, because the questions surrounding expertise are at the core of the effort to simulate human intelligence.

Edward A. Feigenbaum was one of the people from artificial intelligence research who decided, in the mid-1960s, that it was important to know how much a computer program can *know*, and that the best way to learn something about the question would be to try to construct an artificial expert. Joshua Lederberg, the Nobel laureate geneticist, suggested that the task of determining the molecular structure of compounds, based on data from mass spectrography and guided by the rules that are known to govern molecular bonds, was an appropriately difficult and potentially useful problem for artificial intelligence techniques. Together with software expert Bruce Buchanan and Nobel laureate biochemist Carl Djerassi, Lederberg and Feigenbaum started to design DENDRAL, the first expert system, in 1965, at Stanford University.[3]

Human chemists know that the possible spatial arrangement of the molecules that make up any chemical compound depends on a number of basic rules about how different atoms can bond to one another. They also know a lot of facts about different atoms in known compounds. When they make or discover a previously unknown compound, they can gather evidence about the compound by analyzing the substance with a mass spectrograph. The mass spectrograph provides a lot of data, but no clues to what it all means.

Conventional computer-based systems had failed to provide a tool for discovering molecular structures, based on spectrographic data. The problem is that the rules allow a very large number of "near misses"— possible structures that almost, but not quite, fit all the data. There appears to be a "complexity gap" when it comes to the task of sifting through all the near misses. The far simpler computing processes that were used to discover simple structures are just not adequate for more complex structures. DENDRAL was designed to find that one "structure in a haystack" that perfectly fit the spectrographic data and the rules of chemical bonds.

It turns out that you can't just feed all the known facts into a computer and expect to get a coherent answer. That isn't the way human experts make decisions, and apparently that isn't the way you coax a computer into making a decision. What you need is an "inference engine" to fit together the rules of the game, the body of previously known facts, the mass of new data, then venture a guess about what it means.

Building the right kind of "if-then" program, one with enough flexibility to use the kind of rules of thumb that human experts employ, was only the first major problem to be solved. Once you've created a program structure capable of manipulating expert knowledge, you still have to get some knowledge into the system. After feeding the computer program lots of data about molecules, and rules about how they can be combined in molecular structures, the creators of DENDRAL interviewed expert chemists, trying to specify how the experts made their decisions about which combinations and structures are likely to be useful. The resulting program became a milestone in the evolution of software, and the first of a series of software tools for chemists, biologists, and other researchers.

The process of constructing DENDRAL had another useful, unexpected side effect: The task of extracting judgment-related knowledge from human experts led to a new subfield known as "knowledge engineering." "Knowledge engineering" is the art, craft, and science of observing human experts, building models of their expertise, and refining the model until the human experts agree that it works. One of the first spinoffs from MYCIN was EMYCIN—an expert system for those people whose expertise is in building expert systems. By separating the inference engine from the body of factual knowledge, it became possible to produce expert tools for expert-systems builders, thus bootstrapping the state of the art.

While these exotic programs might seem to be distant from the mainstream of research into interactive computer systems, expert-systems research sprouted in the same laboratories that created time-sharing, chess playing programs, Spacewar, and the hacker ethic. DENDRAL had grown out of earlier work at MIT (MAC, actually) on programs for performing higher-level mathematical functions like proving theorems. It became clear, with the success of DENDRAL and MYCIN, that these programs could be useful to people outside the realm of computer science. It also became clear that the kind of nontechnical questions that Weizenbaum and others had raised in regard to AI were going to be raised again when this new subfield became more widely known. As the first frighteningly practical applications to the field of medicine proved when they were created, the field of artificial expertise involves important ethical as well as philosophical, psychological, and engineering considerations.

The clearest area of potential danger in applying knowledge engineering to human medicine is the possibility of misuse through misunderstanding. Although the people who built the system see it as a marvelous but thoroughly fallible tool, many people tend to give too much weight to the recommendation of a computer simply because it comes from a computer. Since medical advice often deals in life and death matters, you have to take into consideration the potential psychological impact of such an "automatic doctor" when you attempt to build something that gives medical advice to an expert.

Like all complex issues, the ethics of medical knowledge engineering have another side. It might be noted by someone from a non-Western, nonindustrial, or nonurban culture that expertise, particularly medical expertise, is a desperately scarce resource. The few medical, hygiene, and agricultural experts who are fighting the biggest humanitarian problems of the world—epidemics and famine—are spread too thin and are working too hard to keep up with scientific progress in their fields. Even in major medical centers, expertise in certain important specialties is a rare commodity.

While so many of the trappings of "modern medicine"—like CAT scanners and other medical imaging technologies—are so expensive as to be limited to a few wealthy or well-insured patients, the potential cost per patient of a software-based system is absurdly low, almost low enough to do some good in a near-future when the number of critically ill people on earth might number in the hundreds of millions.

Medicine—with all its promise and all its difficult ethical implications—appears to be one of the most promising areas of application for commercial knowledge engineering. In the mid-1970s, a physician and computer scientist at Stanford Medical School, Dr. Edward H. Shortliffe, developed MYCIN, the diagnostic system quoted in the earlier dialogue. The problems associated with diagnosing a certain class of brain infections was a technically appropriate area for expert-system research, and an area of particularly pressing human need because the speed with which the infecting agent is identified is critical to successful treatment.

MYCIN's inference engine (the part of the program that makes decisions by applying general rules to specific data), known as E-MYCIN, was used by researchers at Stanford and Pacific Medical Center to produce PUFF, an expert system that assists in diagnosing certain lung disorders. An even newer system, CADUCEUS (formerly known as INTERNIST), uses AI techniques to simulate the diagnostic skills of a specific human physician—Dr. Jack Myers of the School of Medicine at the University of Pittsburgh. Myers and his partner, Harry Pople, Jr., a Carnegie-Mellon-trained AI expert, have been storing parts of Myers' problem-solving style and his knowledge about the entire range of medicine, along with an impressive body of information from the medical literature. CADUCEUS is not yet complete, but it can already perform creditably when it is submitted difficult cases from the medical journals.

Pople told Katherine Fishman, the author of *The Computer Establishment*, that their object is to provide "something the physician would use instead of going to the library or consulting a specialist. There aren't that many experts available, even at major centers." [4] Among the sponsoring agencies who have shown interest in CADUCEUS are NASA, which has an obvious need for such a medical helper in manned space missions, and the Navy, which could use something similar for nuclear submarines. Special gear for astronauts and nuclear submariners might sound remote from most people's daily lives, but in recent history, the transistor radio, handheld calculators, and many other examples of new technologies have traveled from the exotic confines of NASA to the breast pockets of teenagers around the world in less than ten years.

Like the creators of previous technological advances, knowledge engineers first had to prove that expert systems could be built at all and that they were useful. That took about ten years. Next, they had to find potential areas of application—a task that didn't take nearly as long. About two dozen corporations are currently developing and selling expert systems and services. TeKnowledge, founded by Feigenbaum and associates in 1981, was the first. IntelliGenetics is perhaps the most exotic, specializing in expert systems for the genetic engineering industry. Startups in this field tend toward science-fictionoid names—Machine Intelligence Corporation, Computer Thought Corporation, Symbolics, etc. Other companies already established in non-AI areas have entered the field—Xerox, DEC, IBM, Texas Instruments, and Schlumberger among them.

Expert systems are now in commercial and research use in a number of fields. A partial sampling:

- KAS (Knowledge Acquisition System) and TEIRESIAS help knowledge engineers build expert systems.
- ONCOCIN assists physicians in managing complex drug regimens for treating cancer patients.
- MOLGEN helps molecular biologists plan DNA experiments.
- GUIDON is an education expert system that teaches students by correcting answers to technical questions.
- GENESIS assists scientists in planning cloning experiments.
- TATR helps the Air Force plan attacks on enemy airbases.

It's hard to argue with a molybdenum deposit or a significantly high rate of successful diagnoses. As the debate over whether software is capable of acting intelligently dies down in the face of what mathematicians call an "existence proof," the question of whether computer technology *ought* to be applied to such areas as medicine, air traffic control, nuclear power plant operations, or nuclear weapons delivery systems is just beginning.

Some critics, prominent members of the artificial intelligentsia among them, have been sounding alarms over the potential ethical *dangers* of relying too much on electronic artifacts like expert systems to make decisions. Joseph Weizenbaum fears that there is great peril in relying too much on a technology that is very good at *mimicking* what are actually much deeper human thought processes. Expert systems are the epitome of the kind of "imperialism of instrumental reasoning" Weizenbaum rails against—the kind of thinking that sees all problems as solvable through the kind of analytical, mechanical process a computer uses.

In a 1983 interview, Weizenbaum said: "To think that one can take a very wise teacher, for example, and by observing her capture the essence of that person to any significant degree is simply absurd. I'd say people who have that ambition, people who think that it's going to be that easy or possible at all, are simply deluded." [5]

Avron Barr is a knowledge engineer who does not feel that he is deluded, and knowledge-based educational systems happen to be one of his areas of expertise. Surprisingly, Barr agrees with Weizenbaum about the potential ethical dangers of mixing human lives and artificial intelligence research: "Artificial intelligence doesn't exist yet," Barr emphasizes, "but I believe that the kind of research we have started to explore with knowledge-based expert systems can eventually create a tool that truly *understands* human inquiries. And I'm not sure that people are prepared for the ethical decisions that will accompany that kind of power."

From our conversations, and from my perusal of his written work, it became evident to me that Barr also feels that the potential for using this technology to *assist* humanity is well worth pursuing, despite the dangers of misuse. Besides developing and distributing automated expertise to both specialists and ordinary citizens as an informational antitoxin to life in a complicated world, Barr likes to wonder aloud how else might these software entities be used to further positive ends. His personal dream is to eventually build an expert system that is an expert in helping humans reach agreement. If chemists and physicians can use intelligent assistants, why can't diplomats and arms-control negotiators avail themselves of the same assistance? Avron Barr's odyssey through philosophy, psychology, and computer programming has led him to suspect a deep connection between what we *know* individually and how we *agree* collectively.

I first met Avron Barr in a short-order restaurant in the heart of artificial intelligence country—an establishment named "Late for the Train," located next to the Menlo Park train station. If there is an eavesdropping hit list for technological spies, this semiorganic hotcake-and-sprouts joint has to be in the top five. SRI International, one of the oldest robotics research centers, and the birthplace of PROSPECTOR, the molybdenum-sniffing software assistant, is a few shady, tree-lined, affluent blocks away. The tweedy old fellow buttering a scone at the next table looked like a central-casting stereotype of a Nobel laureate.

Barr was wearing a white shirt and tie when we met. He appears to be in his midthirties. His hair is brown and well-groomed, his moustache neatly trimmed—another one of the many babyboomers who might have been hippies in the sixties, but who now go to hairstylists twice a month. He looks like the young man who used to put your groceries in the bag.

Barr got into programming in the first place because he needed a job, and he became involved with artificial intelligence because AI programmers seemed to have the only tools he could find that were capable of helping to create the kind of programs he needed in his work for a research team. His need for a job came after he dropped out of graduate school. His undergraduate work in physics and math at Cornell led to Berkeley, in 1971, where a few months as a physics graduate student made it clear to him that he really didn't want to be a physicist, after all.

At that point, a career in computer science wasn't even on his list of goals, but programming happened to be one of his marketable skills—he had worked his way through Cornell doing scientific programming for various faculty members, stumbling along in FORTRAN, which he taught himself from a book one weekend. After he abandoned his physics career and he began to look for employment, an announcement for a research associate with programming experience came to his attention. The Stanford job called for a resident software handyman in a laboratory that was exploring the technology of instruction. He took it.

He had become a significant contributor to the research team, as well as the hired computer jockey, when he joined a small research group at Stanford's Institute of Mathematical and Social Sciences. Over the next several years, he helped design a program that taught beginners how to program in the language BASIC.

"Which meant I had to go back to thinking about what kinds of people were going to be dealing with computers," Barr recalls, "and finding out what kinds of problems those people might have in the process of learning their first computer language.

"One of the first things that is evident is that computer programs are very different from most of the other things we learn in school because programmers rarely if ever hit the right answer the first time out. Programming is debugging. So being wrong is not so much something to avoid at all costs, but should be seen as a clue to the right way of doing it. That's why it was actually an *environment* rather than just an instructional program. We tried to build a curriculum for teaching BASIC, along with the handholding help people seemed to need in learning software, right into the BASIC language interpreter."

An interpreter, it must be remembered, is not a person who specializes in deciphering computer jargon, but a kind of computer program that can convert programming commands written in the kind of high-level language that people find easier to write into a machine-language form that the computer can read.

The very primitive communications between programmer and interpreter created much of what beginners have always found frustrating about learning old-style programming. Interpreters cannot create programs that will run successfully on computers unless the programs they are interpreting are written perfectly, without a single minor error. If a parenthesis is out of place, the interpreter simply stops operation and puts some spine-chilling message on the screen—the infamous "Fatal Error" or the enigmatic "Syntax Error."

This communication between first-time BASIC programmers and the BASIC interpreter necessary to run their programs was the part of the system Avron Barr and his colleagues were trying to make easier and less frustrating to the human user: "Usually, interpreters return cryptic 'error messages' when they are fed a program with a bug in it," Barr explains. "The program we were building was meant to use the error messages and the debugging as a way to learn how to program."

In order to build an interpreter that not only is able to identify errors, but also can give beginning users hints about how to go about solving the problem, Avron had to go beyond the normal tricks of the programming trade and learn about some of the exotic new notions that were beginning to emerge from AI research. This wasn't standard operating procedure for the vast majority of programmers: To most computer programmers, even scientific programmers, AI was esoteric hocus-pocus that a clique of obsessed academics did with a lot of money from the Defense Department.

When the intelligent interpreter project was finished, Barr entered the computer science department as a graduate student at Stanford, where he encountered Ed Feigenbaum. Although he had been working as a professional programmer, and he was surrounded by artificial intelligence types, and had even picked up a few tricks from AI hackers, this was Barr's first formal exposure to the field. Feigenbaum had an idea about writing and editing a book. Avron took on the task. They thought they could produce a general handbook on AI by the end of the summer. It took five and a half years.[6]

Besides the course requirements of his graduate work, Barr's paying job required him to produce a general text from the contributions of hundreds of AI researchers, a book that someone in a noncomputer-related field could use to get an overview of the most significant work that had been done in AI. The job stretched out longer and longer, and during the time it took to complete his editing duties, he progressed from his master's degree to a Ph.D. in cognitive science.

By the late 1970s, Barr was not alone in feeling that the exploration and engineering of *knowledge*—learning how it is acquired by humans or machines, how it is represented in the mind or in software, how it is communicated between humans and computers and disseminated throughout a culture—was a central problem in philosophy, psychology, and artificial intelligence that might well be answered in surprising ways by the new discipline created by the builders of expert systems.

Computers can track large amounts of information, and they can move through that information very quickly. But when it comes to solving any but the simplest problems—the kind that a human toddler or a chessmaster handle easily—computers run up against a severe problem. Large is never large enough when it comes to the computer memory needed, and fast is never fast enough in terms of computation speed. There is simply too much information in the world to solve problems by checking every possible solution. The difference between brute-force calculation and human knowledge is the missing link (and the holy grail) of hard-core AI research.

Personal knowledge is a tricky thing to describe, and hence a difficult thing to program a computer to emulate. Knowledge is more than a collection of facts, frozen into some rationally coded order. How do our minds do all the things they do when we're thinking, without consciously thinking about how to do it? How do you know which details in a sea of information are worth your attention? The difference between a novice and an expert, for example, is not simply a quantitative question of more stored facts about the area of expertise; the difference hinges, instead, on the ability to make judgments about novel problems in the field.

Chess has been the classic example of the difficulties of emulating expertise with computer programs. It is a finite game, with a limited number of clearly defined allowable moves, each of which have perfectly specified outcomes. Chess qualifies as a formal system in the Turing machine sense, and hence can be imitated by a computer. Give the computer the rules, the starting position, and the opponent's first move, and the computer is capable *in principle*, of calculating all the possible responses to that move and formulating a response based on that calculation.

Yet, after a quarter of a century of effort, nobody has come up with an *unbeatable* chess playing program. The reason that brute-force calculation hasn't defeated a human grandmaster is not rooted so much in technology as in mathematics: the *combinatorial explosion* is the term for the brute-force barrier noted by Shannon back in 1950. Even with only 64 squares and a limited number of allowable moves, the number of possible moves in chess multiplies so quickly that even the fastest computers would take uncountable years to evaluate all legal possibilities.

In chess and many other formal systems, the correct answer is a member of a very large number of possible alternatives. The problem posed by an opponent's move is best answered by a move that will lead to capturing the opponent's king. Hidden among the huge number of possible countermoves for each one of the opponent's moves is one answer or a small group of answers that would have the best chance of achieving the final goal or some intermediate goal. The abstract domain in which the solution is hidden is known as a "problem space."

The brute-force method of finding the right chess move by generating and checking each and every possibility that could exist according to the rules is known as an "exhaustive search of the problem space." Problem space is where the combinatorial explosion lurks, waiting to be triggered by any branching search more than a few levels deep.

The problem of the combinatorial explosion can be easily visualized as a tree structure. If the decisions needed to choose between different moves are seen as the branches of a tree, then a simple two-decision example would yield two branches on the first move, four on the next, eight on the one after that. By the time you get to sixty-four moves, each with twice as many branches as the previous move, you won't be able to see the forest for the branches. If you increase the number of cases to be decided between from two to three, it gets even more snarled: After two moves on a triple-branching tree, there are nine branches (instead of four); after three moves there are twenty-seven (instead of eight), etc., *ad infinitum*. So you have to build in a system to weed out legal but absurd moves, as well as a system to evaluate strategies two or three moves in advance.

What a machine needs to know, practically before it can get started, is that mysterious something that human chessmasters know that enables them to *rule out* all but a few possibilities when they look at a chessboard (or hear a chess situation described to them verbally). When a human contemplates a chess position, that person's brain accomplishes an information processing task of cosmic complexity.

The human brain has obviously found a way to bypass the rules of exhaustive search—a way to beat the numbers involved in searching for solutions in a problem space. This is the vitally important trick that seems to have eluded artificial intelligence program designers from the beginning.

What does the human chessmaster do to prune the tree created by brute-force programs, and how can computers help other humans perform similar tasks? The point of expert-system building is not to outdo the brain but to *help* human reasoning by creating an intelligent buffer between brain processes and the complexities of the world—especially information-related complexities. A problem-pruning tool could be an important component of such an informational intermediary.

Human brains seem to accomplish tasks in ways that would require absurd amounts of computer power if they were to be duplicated by machines. The first expert-systems experiments were not focused exclusively on machine capabilities nor on human abilities, but on the border between the two types of symbol processors. How could a machine be used to transfer expertise from one human to another? The emerging differences between machine capabilities and human cognitive talents were brought into sharper focus when it was demonstrated by systems like MYCIN that this kind of software was capable of measurably augmenting the power of human judgment. Doctors who used MYCIN to aid their diagnostic decision-making ended up making accurate diagnoses more often than they did before they used the program to assist them. The "reasoning" capabilities of the first expert systems were actually quite primitive, but the way these systems worked as "consultation tools" made it clear that there was great potential power in designing software systems that could *interact with people* in ways that stimulated and augmented the exercise of *human* knowing.

The present link between the technology of augmenting human intellect, the business of building expert systems, and the science of artificial intelligence, Avron Barr and his colleagues believe, is the role of *transfer of expertise* both as a practical, valuable tool and as a probe for understanding the nature of understanding: [7]

> A key idea in our current approach to building expert systems is that these programs should not only be able to apply the corpus of expert knowledge to specific problems, but they should also be able to interact with the users just as humans do when they learn, explain, and teach what they know. . . . These *transfer of expertise* (TOE) capabilities were originally necessitated by "human engineering" considerations—the people who build and use our systems needed a variety of "assistance" and "explanation" facilities. However, there is more to the idea of TOE than the implementation of needed user features: These social interactions—learning from experts, explaining one's reasoning, and teaching what one knows—are essential dimensions of human knowledge. They are as fundamental to the nature of intelligence as expert-level problem-solving, and they have changed our ideas about representation and about knowledge.

In order to make a decision with the help of an expert system, a human user must know more than just the facts of the system's recommendation. First, the human has to learn how to communicate with the computer; then he or she needs to know how the system arrived at its conclusion, in terms that he or she can understand. And in order to tell the human about the steps of its reasoning process, such systems must have a means for knowing what they know.

By this point, the exercise has become more than a mechanical search through long lists of possibilities. Problem-solving is only part of the function of a system that must convince a human that the solution it has found is indeed the correct one. The internal and external communication aspects of this transfer process, Barr suspects, offer clues to some of the most significant problems in artificial intelligence as well as intellectual augmentation research: [8]

> We are building systems that take part in the human activity of *transfer of expertise* among experts, practitioners, and students in different kinds of domains. Our problems remain the same as they were

before: We must find good ways to represent knowledge and meta-knowledge, to carry on a dialogue, and to solve problems in the domain. But the guiding principles of our approach and the underlying constraints on our solutions have subtly shifted: Our systems are no longer being designed solely to be expert problem solvers, using vast amounts of encoded knowledge. There are aspects of "knowing" that have so far remained unexplored in AI research: By participation in *human* transfer of expertise, these systems will involve more of the fabric of behavior that is the reason we *ascribe* knowledge and intelligence to people.

Like Doug Engelbart and Alan Kay, Barr feels that future generations will be less inhibited than present-day computer builders and users when it comes to stretching our ideas of what machines and humans can do. This adjustment of human attitudes and computer capabilities is a present-day pragmatic concern of knowledge engineers, and a long-term prerequisite for the kind of human-machine symbiosis predicted by Licklider.

In his conversations, lectures, and writing, Barr often refers to what he and other cognitively oriented computer scientists call "the flight metaphor." Early AI researchers, who were seeking pragmatic means to deal with the question of whether machines could think, compared themselves to those human inventors who not so long ago believed they would eventually build flying machines: "Today, despite our ignorance, we can point to that biological milestone, the thinking brain, in the same spirit as the scientists many hundreds of years ago pointed to the bird as a demonstration in nature that mechanisms heavier than air could fly," wrote Feigenbaum and Feldman in 1963.[9]

"It is instructive to pursue this analogy a bit farther," Barr wrote in 1983:[10]

Flight, as a way of dealing with the environment, takes many forms—from soaring eagles to hovering hummingbirds. If we start to study flight by examining its forms in nature, our initial understanding of what we are studying might involve terms like feathers, wings, weight-to-wing-size ratios, and probably wing-flapping, too. This is the *language* we begin to develop—identifying regularities and making distinctions among the phenomena. But when we start to build flying artifacts, our understanding changes immediately.

Barr then cited another contributor to the history of the flight metaphor, Seymour Papert of MIT, Project MAC, and LOGO fame, who pointed out that the most significant insights into aerodynamics occurred when inventors *stopped* thinking so exclusively about how birds flew. Papert stated to a 1972 European seminar attended by Barr: "Consider how people came to understand how birds fly. Certainly we observed birds. But mainly to recognize certain phenomena. Real understanding of *bird flight* came from understanding *flight*; not birds." [11]

The most difficult barrier faced by the first designers of artificial aviation was not in the environmental obstacles their inventions faced, nor in the nature of the materials and techniques they had available, but in their ideas of what flight could and could not be. The undeniable proof of the simple but incredible idea that flight does not require flapping wings was the most important thing achieved by the Wright brothers.

At the turn of the century, a fundamental part of the problem facing aviation designers lay in abandoning prejudices about the way things actually were, so that the *possible* might be discerned. Those who wanted to build flying machines had to abandon their fixation with the way nature solved the problem of evolving a flying lifeform so that they might see beyond birds to understand the nature of flight. In the same sense, a fundamental part of the problem of artificial intelligence design lies in the ability to see beyond brains or computers to understand something about the nature of intelligence.

Cognitive scientists know that such knowledge can shed light on the way human brains work. Barr points out that such knowledge might expand into varieties of intelligence as different from human intelligence as a jet plane is different from an eagle.

If the flight metaphor could be faithfully extrapolated to the artificers of thinking machines and engineers of programs that understand, Barr claims, new worlds of presently unimaginable information processing mechanisms would become possible—mechanisms that would be compatible with but quite different from the way human brains do things: [12]

> . . . Every new design brings new data about what works and what does not, and clues as to why. Every new contraption tries some different *design alternative* in the space defined by our theory language. And every attempt clarifies our understanding of what it means to fly.
>
> But there is more to the sciences of the artificial than defining the "true nature" of natural phenomena. The exploration of the artifacts themselves, the stiff-winged flying machines, because they are *useful* to society, will naturally extend the exploration of the various points of interface between the technology and society. While nature's explorations of the possibilities is limited by its mutation mechanism,

human inventors will vary every parameter they can think of to produce effects that might be useful—exploring the constraints on the design of their machines from every angle. The space of "flight" phenomena will be populated by examples that nature has not had a chance to try.

Intelligence, like flight, is a way of dealing with the environment. Intelligence, again like flight, conveys a survival advantage to the organism or species that possesses it. The sheer usefulness, the practical value to society, of being able to fly from place to place ensured the development of better artificial ways to fly. Barr suggests that expert systems and other knowledge-based technologies are the kind of "flying machines of the mind" that will have an equally high utilitarian value, and the economics of the marketplace will therefore drive the future exploration of their capabilities.

The "applied" part of "applied AI" is one of the most significant aspects of expert systems, in Barr's opinion, because the linkage of intelligent systems with valuable social goals guarantees the further development of the young science. Because the development of better products in this particular market also means the development of better means of augmenting human intelligence, the evolution of this kind of machine intelligence will be rather closely coupled with the future evolution of human thought: [13]

> It is the goal of those who are involved in the commercial development of expert-systems technology to incorporate that technology into some device that can be sold. But the *environment* in which expert systems operate is our own cognitive environment; it is within this sphere of activity—people solving their problems—that the eventual expert-system products must be found useful. *They will be engineered to our minds*.
>
> . . . It is a long way from the expert systems developed in the research laboratories to any products that fit into people's lives; in fact, it is difficult even to envision what such products will be. Egon Loebner of Hewlett-Packard Laboratories tells of a conversation he had many years ago with Vladimir Zworykin, the inventor of television technology. Loebner asked Zworykin what he had in mind for his invention when he was developing the technology in the 1920s— what kind of product he thought his efforts would produce. The inventor said that he had a very clear idea of the eventual use of TV: He envisioned medical students in the gallery of an operating room getting a clear picture on their TV screens of the details of the operation being conducted below them.

One cannot, at the outset, understand the application of a new technology, because it will find its way into realms of application that do not exist. Loebner has described this process in terms of the *technological niche*, paralleling modern evolution theory. Like the species and their environment, inventions and their applications are co-defined—they constantly evolve together, with niches representing periods of relative stability, into a new reality. . . . Thus, technological inventions change as they are applied to people's needs, and the activities that people undertake change with the availability of new technologies. And as people in industry try to push the new technology toward some profitable niche, they will also explore the nature of the underlying phenomena. Of course, it is not just the scientists and engineers who developed the new technology who are involved in this exploration: Half the job involves finding out what the new capabilities can do for people.

In order to build an expert system, a knowledge engineer needs to encode the rules a human expert uses to make decisions about problems in a specific field, then connect those decision rules with a large collection of facts about that field. The human expert is asked to test the software model. If the human expert disagrees with the system's suggested solution to a problem, then the human asks the system to reconstruct the chain of rules and facts that led to its decision.

By pinpointing the places where the program went wrong, the human expert and the knowledge engineer turn their rough mock-up into a working expert system by a process of progressive debugging. Eventually, they end up with a program that will agree with the human expert a very high proportion of the time. *Consensus* comes in when you ask a second expert to evaluate the system. In real life, human experts disagree with one another, even at the highest levels of expertise. Which means that no matter how well an expert system agrees with one particular human expert, that does not guarantee that another expert won't catch the software making a wrong decision.

The key to taking advantage of these natural disagreements between experts, Barr realized, was to build in a mechanism for "remembering experiences," for keeping around old decisions, even if they were wrong, and creating new rules from the outcome of disagreements. Taken far enough, this aspect of the system leads directly to one of the hottest issues in AI research—the question of whether programs can learn from experience. Barr was only interested in one specific aspect of this issue—the possibility of creating a means of tracking decisions and keeping track of instances where human experts disagree with each other.

"When two experts disagree," Barr explains, "they try to find ways to show each other cases where one or the other's knowledge is not appropriate to produce what they both agree would be the right result. The first steps of establishing consensus, then, involve figuring out where you do agree. Then you can get on to the second step—trying to find exactly where in your individual knowledge systems the disagreement lies.

"Locating the point of disagreement usually turns out to be an important part of the process, because in consciously looking for disagreements the experts realize that they don't share the same meanings for the terms they are using or that they don't share a compatible description of the goal.

"This kind of debugging isn't exciting, but it creates a foundation for the third step of consensus, where the experts have to decide what to do about each other. They can agree that one of them was wrong, they both can remain convinced that they are right, they can decide they are both wrong or both right. They can look for an investigation or experiment that could decide the issue. Or they can decide that they both have to wait for new knowledge."

Barr believes consensus assistance is only a start on "the ultimate kind of thing we can do with intelligent assistants. Consensys [Barr's term for a computerized consensus system] started out as a way of describing how you communicate with one of these systems, in particular, how you might push the expert system to deal with two different human experts and incorporate the value of the differences that the two experts might have.

"My dream has to do with the idea that there is a purpose for us all being here, and we're all necessary for discovering that purpose. Each of us has our own little peephole onto the building being constructed. None of us know what it is, but each of us has a slightly different perspective. And all of those perspectives are necessary to figure out what's being built. It's strange that we can achieve so much as a culture in such a short time, and we can get all these great ideas about how we got here and how the universe works, and yet we know so little about the point of it all. I think that's a clue that computation has a role to play.

"I think of computation as an abstract idea about what it is to share an interpetation of the environment. Computation involves systematic manipulation of symbols, and symbols have a cognitive relation to the world. We need those intermediate messages between our internal representations in order to share perspectives on the world.

"I think it is indeed possible that these kinds of systems will someday be used as a way to work out differences between people. The understanding that is necessary for that to begin to happen involves admitting that we don't know what the purpose is, then finding out why we don't know, and figuring out together how we might come to *understand*. Perhaps computers can play a role in understanding that purpose.

"This might sound very philosophical, but the nature of understanding is at the core of the problems AI programs are up against right now. Pattern recognition in artificial vision or hearing, the ability to understand natural language, the emulation of problem-solving, the design of an intelligent computer interface—all these research questions involve the nature of *understanding*. We don't know what the purpose of understanding is, or why you have to know a whole lot about the world in general to recognize a face or understand a sentence.

"I think most of us believe that understanding is better than not understanding, and that the more we understand the better off we'll be. And I think that the descendants of today's knowledge-based expert systems will help us all to better understanding. Each one of us will be able to understand better because we'll be interacting with people and with information through the assistance of expert tools. They may even help us understand the things that nobody understands."

Few people object to the notion of understanding things that nobody understands—until it is suggested that the agent for achieving that understanding might be an intelligence that is made of silicon rather than protoplasm. The AI infonauts might be on a track that ultimately will bypass the near-future technologies that augment, but do not surpass, human intelligence. If Barr and his colleagues are correct, then their ideas offer strong reinforcement for the speculations that Licklider made in 1960, when he introduced the idea of a coming human-machine symbiosis. Licklider suggested that such a symbiosis was an intermediate step for the interim decades or centuries before the machines surpass our ability to keep up with them.

Even if the human-machine partnership is to be an intermediate relationship, lasting only a few human generations, those next few generations promise to be exciting indeed. When we look at the history of computing, it is clear that even the experts consistently underestimate the rate at which this technology changes. Even the boldest AI pundits might be seriously underestimating the technological changes that will occur in the next fifty or one hundred years.

The paths to the future of mind-augmenting technology appear to be fanning out, the range of alternatives becoming wider and less predictable. It is possible, given past developments, that all of these paths will lead to distinct new technologies, and will precipitate significant changes in human culture. One direction seems to involve the kind of interactive, first-person fantasy amplifiers exemplified by the work of people like Alan Kay and Brenda Laurel. Engelbart's dreams of intellectual augmentation furnish a different model of how the universal tool might evolve. Knowledge engineers like Avron Barr appear to be blazing their own trail. In the next chapter, we'll look at yet another path—one that is more connected to the history of literature than the history of machines.

Ted Nelson, our final infonaut, envisions a future in which the entire population joins the grand conversation of human culture that has heretofore been restricted to those few creators whose works have found their way to library shelves. Wild as his predictions seem to be, they have to be considered seriously, in light of the uncannily accurate forecasts he made back in the "old days" of personal computer history—the 1960s and 1970s.

CHAPTER FOURTEEN

Xanadu, Network Culture, and Beyond

"*Computer* was a bad name for it. It might just as well have been called an *Oogabooga Box*. That way, at least, we could get the fear out in the open and laugh at it." [1]

Ted Nelson is the most outrageous and probably the funniest of the infonauts. Of pronouncements like the one quoted above, he likes to say, "If that sounds wild, it means you understand it"—a statement that could apply to his life as well as his ideas. He's been called "a tin-pot Da Vinci," and "a weirdo who thinks he's a titan"—and that's how *he* describes himself. Opinion in the computer community is mixed when it comes to the question of whether Nelson will ever amount to anything besides a gadfly, pamphleteer, and tinkerer. He seems to have either inspired or irritated most of the key figures in contemporary computing—academic, commercial, or underground.

Even in a crowd of precocious, eccentric loners, Ted seems to set himself apart from the rest. His fate is less certain than those who started augmentation research in the early 1960s or who created the homebrew computer movement in the mid-1970s. Alan Kay is closing in on the marketable version of his fantasy amplifier. Bob Taylor continues to catalyze the development of on-line intellectual communities. Evans

& Sutherland is an extremely successful flight-simulation company, and Ivan Sutherland is a millionaire.

But the idea people in universities and corporate laboratories, the research and development pioneers who made the technology possible, were not the only contemporaries whom Nelson watched and applauded in the mid-1970s as they streaked past him on their way to somewhere. As had happened so often before, some unknown young people appeared from an unexpected quarter to create a new way to use the formerly esoteric machinery. The legend is firmly established by now, and Ted was the first to chronicle it, in *The Home Computer Revolution.*

By the mid-1970s the state of integrated circuitry had reached such a high degree of miniaturization that it was possible to make electronic components thousands of times more complicated than ENIAC—except these machines didn't heat a warehouse to 120 degrees. In fact, they tended to get lost if you dropped them on the rug. In 1971, a team at Intel Corporation developed special integrated circuits that contained all the components needed to make a fairly powerful von Neumann-type computer—the 4004, followed by the 8008 "computer-on-a-chip."

At the time it was invented, nobody realized that the microprocessor, one of thousands of electronic components churned out every year, would become a household word. At that point, probably no more than a few score highly placed or technically fanatic people in the world had computers in their homes for their personal use. IBM and DEC didn't exactly see the invention of the microprocessor chip as the signal to start producing consumer computers.

In 1974, a fellow in New Mexico named Ed Roberts, president of a company called Micro Instrumentation and Telemetry Systems (MITS), happened upon the 8008 chip and got a notion. The chip itself was useless to anybody but an electronic engineer. It had an "instruction set" of "firmware" primitive commands built into it, an arithmetic and logic unit, a clock, temporary storage registers, but no external memory, no input or output devices, no circuitry to connect the components together into a working computer.

Roberts decided to provide the other components and a method for interconnecting them and sell the kits to hobbyists. In January of 1975, *Popular Electronics* magazine did a cover story on "a computer you can build yourself for $420." It was called the *Altair* (after a planet in a *Star Trek* episode). Roberts was hoping for 200 orders in 1975, to keep the enterprise alive, and he received more than that with the first mail after the issue hit the stands.

Bill Gates and Paul Allen were nineteen and twenty-two years old when they wrote a version of BASIC for the Altair. They went to New Mexico to work with MITS, developing software for the first hobbyist computers. It had been obvious from day one that a great many people wanted to have computers of their own. MITS had the usual problems associated with a successful start-up company. Roberts eventually sold it. In 1977, Commodore, Heathkit, and Radio Shack begin marketing personal computers based on the interconnection method established by the Altair—still known as the S100 bus.

Steve Wozniak and Steve Jobs started selling Apples in 1977 and now are firmly established in the annals of Silicon Valley garage-workshop mythology—the Hewlett and Packard of the seventies generation. Gates and Allen became Microsoft, Inc. Their company sold over $50 million worth of software to personal computer users in 1983. Microsoft is aiming for the hundred-million-dollar category, and Gates still has a couple more years before he reaches the age of thirty.

Alan Kay and Bob Taylor and Ivan Sutherland have already been acknowledged and rewarded for their past accomplishments, and look forward to the completion of their future projects under the auspices of well-funded and prestigious organizations. Gates and Allen and Wozniak and Jobs are multimillionaires working on their first billions. They all have what they need to materialize the tools and toys they have dreamed about for decades. Ted Nelson's fortunes, however, have not (yet) turned out so spectacularly.

What Ted Nelson and his long-suffering associate Roger Gregory have now is a long program written in the "C" language—a program that is either a future goldmine for Ted Nelson and boon to all humankind, or yet another crackpot boondoggle on the fringes of computer history. Unsettled as his future might be, what he had in the past was the foresight, the orneriness, and the tenacity to talk clearly and plainly about the computer empire's new clothes.

Ted Nelson was another one of the few people who saw the personal augmentation potential of computers early in the game and grasped the significance of the work being done at Utah, SRI, MIT and PARC. Unlike many of the more sheltered academics, he also saw the potential of a hobbyist "underground." Nelson chose to bypass (and thereby antagonize) both the academic and industrial computerists by appealing directly to the public in a series of self-published tracts that railed against the pronouncements of the programming priesthood.

Nelson's books, *Computer Lib*, *The Home Computer Revolution*, and *Literary Machines*, not only gave the orthodoxy blatant Bronx cheers—they also ventured dozens of predictions about the future of personal computers, many of which turned out to be strikingly accurate, a few of which turned out to be bad guesses.[2]

As a forecaster in a notoriously unpredictable field, Ted Nelson has done better than most—at forecasting. His business and scholarly ventures have yet to meet with success in either the academic establishment or the computer marketplace. He has a history of disenchanting and antagonizing the people who have enough respect for his wild talents to take the risk of hiring him. He's currently on his "third career crash," and still has a while to wait before he knows whether the stock he holds in the company that is going to market his dream will make him a millionaire, thereby vindicating his long struggle, or leave him penniless, thereby branding him as a bona fide crank instead of a late-blooming visionary.

Like so many other computer prodigies, Ted Nelson started his often lonely and always stubbornly unique intellectual journey when he first realized what they were trying to do to him in school. "I hated school all my life," he claims, "from first grade through high school, unrelentingly and every minute. I have never known anyone who hated school as much as I did, although my assumption is that other dropouts do." [3]

Despite his repeated clashes with educational authorities, Ted managed to establish himself as an "extreme loony on campus" at Swarthmore, in the late 1950s, a place and an era when extreme loonies were rather more rare than they became a decade later. He also managed to graduate with an academic record good enough to give him his choice of graduate schools. He decided on Harvard, an institution known to tolerate intellectual arrogance as long as it was accompanied by near-genius originality.

In the fall of 1960, during his second year of graduate school, Ted Nelson found out about computers, and not a moment too soon. He was drowning in his own information by that time, carrying around an already monumental collection of barely collated notes about his abundant dreams and schemes. He found out about Vannevar Bush's paper and embraced the idea that he could use a computer to keep track of his own prodigious stream of thoughts and sketches.

Ted was disappointed to discover that there were no computers equipped or programmed to perform such a service. Down the road at MIT, the first time-sharing computers were only beginning to be built. But Ted needed an information storage and retrieval system to keep track of his notes, and it seemed like such an obvious way to use computers as aids to creative thought that he set out to create such a program himself. Twenty-three years later, he admitted: "It seemed so simple and clear to me then. It still does. But like many beginning computerists, I mistook a clear view for a short distance." [4]

The Harvard course in computer programming that Ted took in 1960 used the only computer then available at Harvard, the IBM 7090 at the Smithsonian Observatory. As a term project, Ted decided to write a machine-language program that would enable him to store his notes and manuscripts in the computer, to change and edit drafts in various ways, and to produce printed final versions. Somewhere around the forty-thousandth line of his program, it dawned on him that his first estimates of the magnitude of the task—and the amount of time it would take to accomplish it—had been overoptimistic.

Nelson's inability to create something even though he was able to clearly envision it is not unusual in the software world. The problem is so widespread that one of the unofficial rules of computer programming (known in some circles as "Babbage's Law") is: "Any large programming project will always take twice as long as you estimate, even if you include that eventuality in your estimate." Even though the simplest of the text-handling capabilities he specified in 1960 were to become, in the hands of other programmers, the software spearhead of office automation in the 1980s, Nelson went far beyond simple text manipulation in the program he set out to write for his term project.

Like Doug Engelbart, whose work he had yet to learn about, Nelson yearned for more than a lazy man's typewriter. They both wanted the freedom to steer their thought paths in new ways. And Ted especially desired the prerogative of changing his mind. He wanted to have the freedom to insert and delete words and move paragraphs around, but he also wanted the computer to *remember* his decision path. One of the specs was for something he called "historical backtrack," in which the computer could quickly show him the various earlier alternative versions of his ever-changing text.

"Alternative versions"? From a place to store notes to a tool for sculpting text, his term project had now landed him in even more wondrous science-fiction territory, a place where it was possible to think in terms of parallel alternatives. Of entire libraries of parallel alternatives, and automated librarians to perform the most tedious of searches in microseconds. Why should we abandon any thought at all? Why not just store every variation on everything and let the computer take care of sifting through it when we want to view something?

Ted Nelson was hooked, and desperately wanted to become a "com-

Ted Nelson—gadfly, prophet, visionary, self-confessed crackpot of the personal computer revolution, 1984. (Photograph by Rita Aero.)

puter person," but came up against the still-prevalent notion that computers are "mathematical." Never one to be accused of excessive modesty regarding his intellectual powers, Nelson admits that he was "a mathematical incompetent." He was even an outsider to those outsiders who were dropping out of MIT and hanging around Building 26. A Swarthmore/ Harvard person just wasn't versed in the way Bronx-Science/MIT people talked about computers.

He couldn't find any jobs as a computer dreamer, but he did manage to find a position as photographer and film editor at a laboratory in Miami where a man named John Lilly was conducting research on dolphin intelligence. Lilly had a very rare piece of instrumentation at that laboratory—one of the original LINC microcomputers designed by Wes Clark. (Nelson didn't use the machine in his work, but its existence convinced him that the idea of small, personal computers was indeed sensible.) After that came a job teaching sociology at Vassar.

Over the next two years, while he taught sociology and thought about the complexities of storing and cross-referencing that had prevented him from finishing his note-keeping program, Nelson realized that he was trying to create a new kind of thing. It was a tool, but it was also a library, and a medium, and a legion of mechanical slave-librarians. In the mid-1960s, when he was working at a book firm, he started to call the whole scheme *Xanadu*. He says it is "a traditional name for a magic place of literary memory," but it is worth noting that Coleridge's poem of that name, like Nelson's term project, was unfinished.

By the late sixties, having offended anyone who could help him in the worlds of academic, commercial, and military computing, Ted was free to find a few like-minded and computer-obsessed friends and attempt to write the software that would make Xanadu possible. By this time, he had not only dreamed up the specifications for the full-blown version of this new information processing system, he had managed to attract a few equally fanatic allies.

The basic note-keeping scheme that started it all was meant to have a system for automatically taking care of backtracking. The next step was to expand this capability to handle alternative versions and to show the user which parts of different versions are the same and which are different. This is the *versioning* capability, which Nelson now estimates to consume about 5 percent of the Gross National Product—from the boiler-plate paragraphs used by attorneys to the 47 different versions of the 747 design that are stored in Boeing's computers. In real life, there is hardly ever such a thing as "the contract" or "the 747 blueprint." Mixtures of standard and custom features that make for slightly different versions of contracts or blueprints are more often the case.

Historical tracing and versioning, however, don't make for much more than a powerful word processing system. Things started getting extradimensional when Nelson thought about adding *links*. Engelbart thinks that he and Nelson just happened to come up with something similar around the same time, although Engelbart had the technology and the wherewithal to actually get such a system up and running. The whole idea started out as a kind of computer-dynamized footnote—a way to jump from part of the text to something outside the main body of the current document.

Instead of encountering an asterisk and looking down at the bottom of the page for a footnote, and possibly looking up another document elsewhere in the library to verify a reference, the user would point a lightpen or a mouse at the electronic equivalent of the asterisk, and automatically bring the appended or referenced material to the screen. A return button would bring the user back to the point in the original text where the link symbol appeared. A very similar feature was built into Doug Engelbart's early NLS system.

Engelbart was more concerned with constructing the toolkit and workshop for solving problems than speculating about the kind of literary form such a facility might create. Nelson, however, being a liberal arts type rather than an engineering type—a dichotomy he deplores, since it kept him away from computers for so long—wondered what art forms and intellectual systems might emerge. In its simplest essence, a link is a reminder to the reader that "there is something to jump to here." Links meant that literature no longer had to be sequential.

The link facility, Nelson insisted from the first, provides something far more powerful than a means of attaching odds and ends. A system with backtrack, versioning, and links would create the possibility of a new way of organizing thoughts into words, a nonsequential form of writing that was never possible before computers, a literary process he called *hypertext*.

Hypertext, as he first imagined it, could apply to scholarship as well as to poetry. Scientific literature, the very basis of worldwide scientific scholarship, consists of published documents which refer to many previous documents. An experiment is usually performed to test a hypothesis that was based on previous experiments. Performing a "search of the literature" is the first thing a scientist does when confronted with a new research problem.

The problem today is that scientific research is *too* successful. As Vannevar Bush warned forty years ago, the rate and volume of scientific publication have overwhelmed the coping capacity of our old print-era technology. With a hypertext system, each scientific document could have links to its intellectual antecedents and to documents regarding related problems. The entire body of relevant scientific literature could be collapsed into each individual document. The links would function in the same way as footnotes, but with immediate access to the cited material, as if each footnote was like a window or door into the cited document.

A system with links, backtrack, and versioning needs only an economic structure to become a publishing system. Nelson sees an anarchic but self-organizing system based on his conception of royalties and subroyalties. In a Xanadu-like system, royalties are automatically monitored by the host computer network, and are based largely on transmission time— the amount of time people pay on-line attention to a given document. Every document in the system has an owner, and every owner is paid "a whiff of royalty" whenever somebody calls their document from the memory and displays it in words, sounds, or images.

Everybody can create what they want and put it on the system, from sonnets to pamphlets to textbooks, and everybody can quote or cite any other document. Documents can consist of links. Compendia, guided tours, directories, and indexes will spring up as independent documents; order would become a valuable commodity. "The result is a seemingly anarchic pool of documents, true, but that's what literature has been anyhow . . . , " Nelson claims. "Its orderliness is not, as some would suppose, imposed by the computer or its administrators, but by something which arose long ago in the natural structure of literature, and which we are merely retaining." [5] Just as literary critics and librarians have found ways to organize and categorize the apparently chaotic stream of traditional literature, Nelson claims that people will spontaneously invent methods of organizing a hypertext-based body of literature.

Nelson sees his ultimate concerns about the technology as political. Where most revolutionaries have regarded the computer as a tool of totalitarian oppression, a symbol of centralized power and dehumanization, Nelson has long known that these ideas are based on an outmoded kind of computer. Distributed networks of individually powerful computers are an entirely different thing from a central computer with a lot of extensions, and Nelson was one of the first to point out this technology's potential for creating social forms directed by the individual members, who are beyond the command of any old-fashioned, mainframe-type central control. He is enthused by the personal power that comes with having ready access to usable forms of information—the bite of the old hacker apple—and zealous about preserving the freedom to explore it in your own way: [6]

> Those of us who grew up believing passionately in ideals that made
> our country great, such as liberty and pluralism and the accessibility

of ideas, can hardly ignore the hope of such an opening-out. Libertarian ideals of accessibility and excitement might unseat the video narcosis that now sits on our land like a fog. I want to see the writings of Herodotus, Nostradamus, and Matthew Brann as accessible as those of Rod McKuen, along with the art of the Renaissance and movies of tomorrow—an all-encompassing picture-book encyclopedia tumult graffiti-land, the Whole Works.

If this all seems like a wild idea, that means you understand it. These are times wild with possibility. In an age of pocket calculators, the Pill, hydrogen bombs by rocket, and soap operas by satellite, we can try to create whatever wildness we want in our society.

. . . I say these worlds are possible soon. We need them, and they will make lots of money. The software is on its way. But what is really lacking are the visionary artists, writers, publishers, and investors who can see the possibilities and help carry such ideas into reality.

What Nelson is raving about is not a technology, but a *community*. The idea of electronic communities is no longer just an idea. Lap-sized computers with crude display screens and built-in modems are already on their way to becoming commonplace. The visual displays will grow far more sophisticated, and the computers' processing power will increase as prices drop. Dynabooks and ARPAnets are suddenly not limited to research laboratories or military bureaucracies. On-line interactive communities are evolving right now, all around the world, through the wholly voluntary efforts of teenagers with modems, traveling business people with briefcase telecomputers, information utilities, computer bulletin board systems, and telecommunes of every stripe.

Ted Nelson is voicing what a few people have known for a while, from the technical side—that the intersection of communication and computer technologies will create a new communication medium with great possibilities. But he notes that the art of showing us those possibilities might belong to a different breed of thinker, people with different kinds of motivations and skills than the people who invented the technology. After Gutenberg came Cervantes. After movable type came novels. As Alan Kay pointed out, literature was the software of the print era. The Cervantes of Hypertext might be learning to read right about now.

Twenty years ago, the few hundred people who built time-sharing began to get excited about several new means of communication that were becoming possible via computer mediation. Fifteen years ago, the thousand-odd people who joined the first version of ARPAnet began to experiment with these new media—in their daily work and as a way to have fun. About a decade ago, another group of people began to concentrate on software systems specifically designed to facilitate communications among a dispersed community—*computer teleconferencing*.

The concept of computerized conferencing came from the usual convergence of unexpected factors—in this case, the Berlin airlift of 1948, a decision tool invented by a think tank, and the wage-prize freeze of 1971. The idea was to build a system in which computers make it possible for groups of people who are separated by both space and time to communicate in various ways, over common-carrier communication lines. Community communication was first tried during the Berlin airlift, when the only agency with direct real-time communications of its own to all the NATO countries was the State Department, with its old-style teletype machines. Somebody tried to wire all the machines together, without the aid of computers to help organize the message-stream—which created a classic mess, and the classic story of the birth of the new medium.

The earliest development of the idea of using computer mediation in geographically dispersed conferences is most widely associated with Murray Turoff, the standard eccentric prodigy, the character who happens to see everything differently and who, like the other young, independent-minded thinkers before him, liked to follow an idea wherever it led him.

In the late 1960s, Turoff was working on war games and other kinds of computer-based simulations for a Washington, D.C., think tank, the Institute for Defense Analysis. Some of these games involved connecting several "players" at once, via remote computing systems. As a result of this experience, Turoff became interested in using computers to mediate a special process developed at Rand, known as the "Delphi method," in which printed questionnaires and responses circulate among a community of experts. Delphi was a way to reach a quick collective judgment about a complex situation; Turoff thought the process was ideally suited to the kind of on-line communications then being demonstrated on the ARPAnet. So he started to experiment with a computerized Delphi system.

In the early 1970s, Turoff had moved to the Office of Emergency Preparedness, where his job wasn't related to his immediate interests in teleconferencing. His superiors found out that he was using his computer terminal to experiment with an unauthorized conference system, and there was some on-the-job friction. But then came the wage-prize freeze of 1971, an action that required the rapid collection and collation of an unprecedented amount of information. Turoff's superiors changed their minds. The Delphi Conference System was ready for action just in time.

In the process of putting it together, the people who designed the system and the people who used it began to discover that some of the system's features just seemed to become popular with the on-line community, with no official urging and often with no connection to the task at hand. There was, for example, a feature they simply called "messages." Anyone plugged into the system could leave a message for anyone else on a kind of computerized blackboard. Like a blackboard, you could check your message later and see if anyone had appended a note. Notes proliferated so fast that people began to develop programs for sifting through them.

The fancy part of the software came in when you wanted to be able to review only the last five messages, or only those relating to a particular topic, or all the messages from a particular person, or on a given date. Similar efforts to build electronic mail systems were also going on in conjunction with the ARPAnet. One unique feature of both systems that emerged early was the capability of communicating with a specialized audience, even if you didn't know who was in that audience. For example, if you indicate to the host computer that you want all future messages on the topic of AI research, folk dancing, and Spacewar to be routed to your electronic mailbox, then anyone with news about one of those topics can reach you without knowing who you are.

They were all discovering something that had been unknown in previous communication media—the content of the message is capable of also being an address. Far from being a tool of dehumanization, the computer conferencing system could boost everybody's ability to contact a community of common interest. Some kinds of teleconferencing software were created in order to make it possible to post a message on the topic of zucchini or microprocessors (or emergency preparedness procedures, or organizing an airlift) and be sure that the message would be transmitted to everyone who needed to know about those topics.

The use of a computer-mediated message system, as Turoff understood, ultimately created several new *social* phenomena. It was obvious from the vigorous electronic mail traffic on the ARPAnet that some new kind of conversation was going on. At a technical level, the users of these systems were able to share computer resources and research findings, as they were supposed to. But it also turned out that whenever people are introduced to a computer network, they seem to want to use it to *communicate* with each other.

People on the ARPAnet devoted hours to composing messages. For the small community of people who had access to such systems, the continuing dialogues on AI and foreign policy, space shuttles and Space-

war, diatribes, puns, puzzles, gossip, pranks, and running jokes became
a kind of combination electronic water-cooler and customized daily news
medium. All the other news media were collapsed into subsets of the
new one, since it was no problem to plug the wire services into the
system. The metamedium seemed to foster new kinds of values, as well.
Iconoclasm, debate, the right to an unbridled heterogeneity of interests
seemed to be highly valued in the emerging on-line community.

In some quarters of that community, people like Turoff and Engelbart
were trying to learn enough from network communication behavior to
help them design new tools for group communications. The National
Science Foundation, deeply concerned with the problem of establishing
a new way for the half-million scientists in this country to communicate
with each other, supported some of the conferencing research. Under
NSF sponsorship, Turoff moved to the New Jersey Institute of Technology (NJIT), to both study and improve the technology. A similar project
had already begun in California, at a place in Menlo Park, not far from
SRI and PARC, called the Institute for the Future.

Roy Amara and Jacques Vallee and other staff members at the Institute
for the Future worked on a system known as PLANET (for Planning
Network, because it was initially directed at planners in government
and industry). Both Turoff's and the institute's systems began with electronic mail, a shared notebook space for joint compositions, a conference
facility for on-line and off-line group communications, and an open-
message/bulletin board.

Turoff and his associates' EMISARI system that had evolved from
the Delphi Conference System evolved again into the RIMS (Resource
Interruption Monitoring System) which has been used, according to
Turoff, by the "Federal Preparedness Agency in every major national
commodity shortage and transportation strike since 1971." [7]

But by the time he joined NJIT, Turoff's interest had expanded beyond
the development of a communications tool for crisis management: "I
think the ultimate possibility of computerized conferencing is to provide
a way for human groups to exercise a 'collective intelligence' capability,"
he noted in 1976. "The computer as a device to allow a human group
to exhibit collective intelligence is a rather new concept. In principle,
a group, if successful, would exhibit an intelligence higher than any
member. Over the next decades, attempts to design computerized conferencing structures that allow a group to treat a particular complex problem
with a single collective brain may well promise more benefit for mankind
than all the artificial intelligence work to date." [8]

In 1977, the National Science Foundation funded the NJIT to build "an electronic communication laboratory for use by geographically dispersed research communities." [9] By July, 1978, seven trial projects were under way, each one a part of an established research community of ten to fifty members. The system was set up to collect data on its own operations, in order to test the hypothesis that a teleconference-like system could enhance the effectiveness of research communities.

The Electronic Information Exchange System, known as EIES (pronounced "eyes"), was another one of those experiments that never shut itself down because the experimental subjects just wouldn't let go of it. It seemed to happen with every new development of interactive computing—people would simply refuse to stop experimenting with the system, and wouldn't give up the experimental tools when the experiment was over. As Jim Fadiman noted of ARC, people seem to be as reluctant to be deaugmented as they are resistant to augmentation in the first place.

EIES was first set up to enable members to send private communications to individuals or groups, maintain permanent transcripts of comments on discussion topics, and provide text processing and file management services that participants could use to construct jointly authored papers. The protocols for using all the communication features, like Engelbart's NLS system, were not easy to learn. It took some commitment to the idea that it was worthwhile learning, which is one reason why research communities were ideal laboratories for the experiment.

EIES quickly expanded from pure scientific research communities to legislative researchers and medical researchers. Another project in the late seventies used a modification of Engelbart's NLS system to enable EIES subscribers in one experimental community to quickly browse through time-sensitive technical information. By 1978, policy-makers, artists, long-range planners, and others began to join EIES. Roxanne Hiltz and Turoff published a book that year, entitled *Network Nation*, in which they predicted that the medium wouldn't be limited to a few laboratories and think tanks. They noted that any microcomputer with a modem and appropriate software could plug into any network its user knew how to enter. They saw the development of easier-to-use, population-wide teleconferencing networks as a means of reducing the distance between people's minds and thoughts, as a forum for intellectual discourse and group decision-making, and as a model for a new kind of community where one's age, gender, race, or physical appearance would no longer matter as much as what one has to say.

By the early 1980s, personal computers were being sold by the millions, and some of the people who bought them wanted to plug into these networks they were beginning to hear about. EIES has always been something of an elite—you have to apply and pay a relatively high fee. But the first public information utility wasn't long in coming. In June, 1979, the Telecomputing Corporation of America opened for business out of a host computer in McLean, Virginia. Reader's Digest bought the company in 1980 and it was renamed Source Telecomputing Corporation. Reader's Digest, not an organization known for small-scale pursuits, carried the organization through the early years when computer sales crept into the hundreds of thousands. By the end of 1982, The Source had over 25,000 subscribers, and a growth rate of over 1000 new subscribers per month. Satellites and state-of-the-art computers and new software were added to accommodate up to a quarter-million subscribers.

To those who can afford an initiation fee of $100, and a connect-time fee of $7 to $22 per hour, The Source and its newer competitor, Compuserve, offer computer owners admission to an electronic community-in-the-making. Besides remote computing, electronic mail, communications, conferencing, telemarketing, software exchange, game playing, news gathering, bulletin board, and other services, The Source provides something called "user publishing."

Since subscribers are billed according to how much time they spend with their computer connected to the Source host computer, it is possible to pay royalties to "information providers," based on a portion of that connect time. Every time a Source subscriber reads wire service information, the information provider gets a cut of the take. The same is true of user publishers. You have to pay for everything you put in storage, so the popularity of your service with the subscribers is what determines whether any publication is economically viable. To a creative writer, the challenge is tempting—as long as you can keep your audience reading, the royalties will outweigh the storage charges. The artist can now be the publisher and go directly to the audience.

Two electronic magazines I encountered my first time out were called *Sourcetrek* and *Mylar's Warp*. *Sourcetrek*, subtitled "Journeys through the Electronic Void," is put out by "Sourcetronaut Dave," aka "Sourcevoid Dave." When you give The Source the command to connect you to *Sourcetrek*, you get a menu of choices on your screen, along with a list of different statistics about the choices—reading time, number of times read, the exact time it was last read. I selected the first "article," entitled "Hello," which went (in part) as follows:

Hello.
I am "Sourcevoid" Dave. David Hughes otherwise.

I was born in Colorado, descended from stubborn Welshmen who were never too loyal to the king. Which is probably why I am content being a maverick of sorts, with a Welsh imagination.

I live in Historic Old Colorado City at the base of 14,114 foot Pike's Peak.

I work out of my 1894 Electronic Cottage with a variety of micro-computer and telecommunications tools. . . .

I am a happily married middle-aged family man who has seen enough of Big Government, Big Wars, Big Industry, Big Political Causes—either of the left or right—to now prefer to operate a small business out of a small house, in a small neighborhood, working with small organizations, using a small computer to make it all possible.

I also have a small computer bulletin-board to link my local friends with my brain—asynchronously and in the noble written form of English. . . .

Dave has opinions and poems and stories to tell. He teaches classes via modem to students around the world. And all subscribers can read what he has to say, at their own expense, and reply by electronic mail if they wish, also at their own expense. The other electronic magazine I sampled, *Mylar's Warp, an Electronic Serial*, by Floyd Flanagan, was strictly fictional. The idea is the same idea behind any serial—the writer has to keep it interesting in order to keep the readers' attention.

The title of Chapter 1 was "Reflections on Ice," and this is as much as I read before I realized how much I was spending in connect time:

I know I'm freezing to death. Wasn't supposed to feel a thing. Ha! A sucker born every minute. Just because you're frozen alive, that doesn't mean you can't still be freezing to death. I may be slowed down but I ain't dumb. Sure as hell, I'm freezing to death.

So, how did I get here? No reason not to go over it again for the eleven millionth time. Nothing else to do. I'm Johnny Mylar, from Peabody, Utah. Peabody's claim to fame was Dinah, a life-size pea-green, cement replica of a dinosaur, like me, frozen out of time. . . .

Anyway, it all started when I was getting my driver's license renewed and the lady asked me if I would like an organ donor sticker on the back of my license. Hadn't really ever thought about it before, I told her. So, she explained how, if I died and there was a sticker on my license, then the hospitals would be able to use my organs to help people who had lost an eye, or heart, or brain, or tooth, or whatever. "Sure," I said. "Whatever's right." I had always had a cavalier attitude concerning the most basic matters, like sex and death. Didn't I always buy Girl Scout cookies from the little girls in the short green skirts, and . . . "

While the community of subscribers to EIES, The Source, Compu-serve, Dow Jones, and other information utilities is still small enough to keep the costs of the service high, the inevitable growth of the telecom-puting population from tens of thousands to millions, spurred by the proliferation of modem-equipped home computers, is sure to lower the price enough to make it possible for more Floyd Flanagans and David Hugheses to experiment with their electronic magazines. But the big info-utilities are not only the kind of on-line community in existence. At the same time that the larger utilities seek to plug individual subscrib-ers together into what is essentially a centrally controlled time-sharing technology, a different way of interconnecting computers is giving birth to an even wilder mutant of network culture—the computer-based bulle-tin boards.

A computer bulletin board system, often called a CBBS, or simply a BBS, consists of a computer controlled by special software and the hard-ware needed to connect it to an ordinary telephone line. The software enables a small host computer to automatically answer when its number is dialed, then transmit and receive messages to and from remote comput-ers. By leaving such a system hooked up continuously, and posting the access number in one or two places, the grapevine takes care of the rest. Come back and read the messages a week later and you'll discover that a community has created itself.

The first software that enabled microcomputer owners to set up CBBS was created by Ward Christensen and Randy Seuss, in Chicago, in 1978. By 1984, the number of such systems is difficult to determine, but must be at least in the hundreds, and probably will soon be in the thousands. To connect to a BBS, you need a personal computer, a modem, telecommunication software, and a telephone. Plug the telephone into the modem, use the communication program to dial the BBS number, then when the computers are connected, the host system will put words on your screen and tell you how to work the system.

Most people know of these systems, and the underground community of users, because of the movie *WarGames*, television programs about computer whiz kids, and publicity about dark-side hackers. In fact, the community has changed so swiftly that piracy, phone-freaking, destructive hacking, and even obsessive interest in how computers work now occupy only a small part of the BBS scene. Many bulletin boards have been in existence for years, but even more seem to spring up and die out on a weekly basis. In my own limited sampling of the BBS world, over the span of a few months, I encountered teenage philosophers, homespun lecturers of all ages and both sexes who were willing to ramble about any topic you'd care to name, and I even stumbled onto a couple of on-line religions, both cybernetic and pagan.

I met Clyde Ghost Monster one night out in the bulletin board zone, and Clyde ultimately turned me on to the number that led me to the on-line religion. It started the way it usually does when you browse the boards. A list of bulletin board numbers had led me to a list of bulletin board numbers that led me to another lively discussion group called "Sunrise" in New Jersey, consisting of random drop-ins from anywhere in the country, like me, and a core group, mostly local, who seemed to know each other, and who spent hours trading messages about utterly anything at all.

While some boards are strictly for hackers or computer enthusiasts or science-fiction freaks or sex freaks or peace types, Sunrise appeared to be a kind of electronic cracker-barrel store crossed with a public restroom wall. I joined Sunrise as "Johnny Jupiter" when I decided to add my two cents to a very funny ongoing conference that consisted of nothing but lists of "my favorite people." You can say a lot with just a list of people, the Sunrise community discovered one night, when "Ivan Idea" started it all by posting the first list. The creators of the lists that followed within hours signed themselves with names like "Tater Tot," "Clock Speed," and "Clyde Ghost Monster."

I checked in on Sunrise from week to week, and one night, while scrawling some graffitist reply to an ongoing epistemological debate, the words "SYSOP REQUESTS CHAT" appeared on my screen. I typed "OKAY LET'S CHAT," hit the return key, and started conversing in real time with an utterly fascinating individual, via an exchange of quickly typed messages.

It turned out that the host computer was located in Clyde Ghost Monster's bedroom, which made Clyde the system operator. Sysops are like benevolent dictators. They can weed you out of the community memory if they want, but then again, their computer is the one that provides a message-mediation service to anyone who wants to drop in, electronically speaking. Clyde Ghost Monster was an anarchist sysop, who preferred the rule of wit. Clyde Ghost Monster, I was to learn, weeks later, was also a sixteen-year-old girl. Tater Tot was a seventeen-year-old boy who went to her high school. They had no idea who Ivan Idea was.

Clyde told me that if I wanted to find out about new kinds of communities, I ought to call a conference-tree bulletin board in Santa Cruz, California, and read the opening message for "ORIGINS." The conference tree is a bulletin-board-based medium that seems particularly well suited to wildly heterogeneous experiments in communitarian communication. The idea behind a conference tree is that you can call in and read from or write to a variety of conferences, each one consisting of a constantly branching list of messages and submessages. The name of the message conveys something of what it is about, and all the variations of opinion from rabid enthusiasm to utter contempt can be expressed in submessages and submessages of submessages.

My modem beeped its way to the host computer, and when the word CONNECT appeared on my screen, I hit the return key twice. A menu of conferences appeared, in the form of the list of names of the first message in each conference. I selected "ORIGINS," as instructed. ORIGINS first gave me an address to write to obtain a brochure, then the following message appeared on my screen:

> ORIGINS is a movement that started on this computer (Santa Cruz, 408-475-7101). ORIGINS began on the START-A-RELIGION conference, but we don't call it a religion.
>
> ORIGINS is partly a religion, partly like a Westernized form of yoga society, partly a peace movement. It is a framework for improving your life and improving the world at the same time.
>
> The movement centers on "practices"—actions you can use in everyday life to build effective human relationships, strength of community, and self-awareness. All the practices are based on action. None require any special equipment, settings, leaders, theories, or social status. The human universals of the ordinary, everyday moment, and the personal relationship, form the basis for this training.
>
> ORIGINS has no leaders, no official existence, nothing for sale. Because it started in an open computer conference, no one knows who all the creators are.
>
> This movement has just begun. The brochure mentioned above recommends seven practices (Leverage a favor, Ask for help and get it, Use charisma, Finish a job, Use magic, Observe yourself, Share Grace), but these suggestions are only starters. The idea is to continually develop new training/action methods, as a community project, then discuss and share them through whatever communications media are available. This movement will never be finished, because it seeks a community of permanent innovation.
>
> The hope is to build something which can make a better world. The first step is to make your own life better. For a more detailed overview of ORIGINS, get the brochure from the address above. To see how the movement developed, read the START-A-RELIGION message and its submessages.

Although the conference tree that contained ORIGINS, along with its parent and sister and daughter conferences (as submessages and root messages are known in BBS jargon), was one of the most intriguing electronic gathering places I found in a few months of vicarious wanderings via my modem, it was far from the only unusual one.

The pros and cons of religion, and the possibility of starting new ones or reviving old ones, seems to be a popular topic of discussion. ORIGINS was an example of the cybernetic variety. I ran across a few Christian boards and a meditators' BBS, but the most startling discovery in my board-browsing was the pagan faction who announced themselves with a message on a conference tree:

> The Covenant of the Goddess is an umbrella organization for pagan groups of all kinds. It was created in the 60s to provide some structure (and maybe some muscle, since some groups were being harassed by the government) to an otherwise amorphous bunch of covens in Northern California, but eventually had members everywhere. A pagan group mostly refers to witches, although there are Druid groves and other strictly unallied organizations online as well. Witches means any affinity group which holds as one of its general tenets that Jehovah may not be the guy in charge after all—that he is a powerful illusion created by an awful lot of misguided and power-hungry folks, and that the supreme being is and should be somebody with more of a sense of humor as well as compassion, not even to mention love. In short, it might be fair to say that if it doesn't claim that it's better than any other way, then it's probably pagan. These definitions are by exclusion because one way of defining the whole pagan movement is as a group that believes in saying yes to more. A coven is an affinity group of witches. The name is very old. Some covens have fierce strict codes of behavior and rules of ceremony and others get together now and then and shoot the shit. By and large, witches have the best parties of any religious group going. There is another organization in the California area known as the New Reformed Orthodox Order of the Golden Dawn, which was started as a gag in the 60s and presently has several thousand members, a good many of which can apparently be counted on to show up for a bash. It is typically pagan, incidentally, to start your biggest umbrella organization for a joke. Lots of witches compute, and there are probably a bunch on this very tree who have not bothered to identify themselves. (Witches have no identifying marks—except that humorous glint in the eye.)

Religion, ancient or modern, is still less popular than sex as a topic for BBS discussion. A certain steady percentage of boards are entirely sexually oriented. The problem used to be that there simply weren't any females in the system, a situation that appears to be changing rapidly. Sexually oriented CBBS and dial-a-date boards are an entrance into yet another subcult within a subculture, some members of which use the system to arrange real-life assignations with compatible companions, but most of whom use the system to live out fantasy sex lives consisting of hot dialogues with other anonymous participants.

Because computer programs can be sent over the telephone wires as easily as words or numbers, some boards engage in software piracy— passing along proprietary software without paying a licensing fee. Others dispense "public domain" software as a community service. Some of them offer access to special information, like an insider newsletter, and issue passwords and bill for connect-time. Some are exclusive, and many are promiscuous, about who is allowed to write as well as read messages.

Then there are the folks who are starting to use *temporary* on-line communities as art forms and experiments in changing the consciousness of larger communities like neighborhoods and cities. In 1983, a literary group in Seattle that called itself *Invisible Seattle* instigated the creation of a fifteen-chapter mystery story written by a representative sampling of the half-million citizens of the city itself. The collective novel was not a new form, as far as the more standard kind of networks go. EIES started a serial years before, in which different writers took on the personae of various characters and wrote the story like a conference.

Invisible Seattle, however, sent "literary construction workers" out into the city looking for people from all walks of life who were willing to contribute plots, words, ideas, which were communicated from the point of origin to the other nodes throughout the city via a temporary arrangement of video arcade game parts, two larger personal computers, some custom-written software, and six smaller personal computers.

What do Xanadu, EIES, The Source, Clyde Ghost Monster, and Invisible Seattle have to do with the technology created by Turing, von Neumann, Licklider, et al.? What would the patriarchs think of the infonauts? The changes that were predicted by the earliest software prophets seem to be only beginning. The religion that germinated on the ORIGINS conference tree—was its origin any stranger or less likely than the dominant religions of today that sprang up centuries ago in dusty Middle Eastern villages? Xanadu and EIES might seem like novel and unfamiliar media—but so did printing presses and telephones when they first appeared.

The forms that cultural innovations took in the past can help us try to forecast the future—but the forms of the past can only give a glimpse, not a detailed picture, of what will be. The developments that seem the most important to contemporaries, like blimps and telegraphs, become humorous anachronisms to their grandchildren. As soon as something looks like a good model for predicting the way life is going to be from now on, the unexpected happens. The lesson, if anything, is that we should get used to expecting the unexpected.

We seem to be experiencing one of those rare pivotal times between epochs, before a new social order emerges, when a great many experiments briefly flourish. If the experiences of past generations are to furnish any guidance, the best attitude to adopt might have less to do with picking the most likely successors to today's institutions than with encouraging an atmosphere of experimentation. Is Ted Nelson any crazier than Alan Turing? Did Gutenberg think about the effects of public libraries?

Hints to the shape of the emerging order can be gleaned from the uses people are beginning to think up for computers and networks. But it is a little bit like watching the old films of flying machines of the early twentieth century, the kind that get a lot of laughs whenever they are shown to modern audiences because some of the spiral-winged or twelve-winged jobs look so ridiculous from the perspective of the jet age. Yet everyone can see how very close the spiral-winged contraption had come to the principle of the helicopter.

The dispersal of powerful computer technology to large segments of the world's population, and the phasing-in of the comprehensive information-processing global nervous system that seems to be abuilding, are already propelling us toward a social transformation that we know very little about, except that it will be far different from previous transformations because the tool that will trigger the change is so different from previous tools. Not all of those who have tried to predict the course of this transformation have been as optimistic as Licklider or Nelson. Joseph Weizenbaum, in particular, has voiced his fear of the danger of mistaking computers for human minds or treating human beings as machines.

Weizenbaum's argument, in part, points out that the aspect of human nature that was externalized by the invention and evolution of computers was precisely the most machine-like aspect. The machines that embody this aspect can do some very impressive things that humans cannot do, and at present can do very little of the more sophisticated intellectual feats humans can accomplish. Even so, they are taking over management of our civilization. Before we begin to give more of our decision-making responsibility over to the machines, Weizenbaum warns that it is a terrible mistake to believe that all human problems and all important aspects of human life are computable.

This "tyranny of instrumental reasoning" can lead to atrocities, Weizenbaum warns, and in the closing years of the twentieth century, it is not at all paranoid to have some healthy suspicions about what any shiny new technology that came from the Defense Department in the first place might do to our lives when they get around to mass producing it. And there is no dispute that war was the original motivation and has been the continuing source of support for the development of computer technology.

If it is true that the human brain probably started out as a rock-throwing variation on the standard hominid model, it has also proved capable of creating the Sermon on the Mount, the *Mona Lisa*, and *The Art of the Fugue*. If it is true that the personal computer started out as an aid to ballistic calculations, it is also true that a population equipped with low-cost, high-power computers and access to self-organizing distributed networks has in its hands a potentially powerful defense against any centrally organized technological tyranny.

Licklider believed that a human-computer symbiosis would be the means of steering our planet through the dangerous decades ahead. Others have used another biological metaphor for our future relationship with information processing technology—the concept of *coevolution*, an agreement between two different organisms to change together, to interact in such a way that improvements in the chances for survival of one species can lead to improvements in the chances for survival of the other species.

Perhaps yet another biological metaphor can help us foresee the transformation ahead. When a caterpillar transforms into a butterfly, it undergoes a biologically unique process. Ancient observers noted the similarity between the changes undergone by a butterfly pupa and those of the human mind when it undergoes the kind of transformation associated with a radical new way of understanding the world—in fact, the Greek word for both butterfly and soul is *psyche*.

After the caterpillar has wound itself with silk, extraordinary changes begin to happen within its body. Certain cells, known to biologists as *imaginal cells*, begin to behave very differently from normal caterpillar cells. Soon, these unusual cells begin to affect cells in their immediate vicinity. The imaginal cells begin to grow into colonies throughout the body of the transforming pupa. Then, as the caterpillar cells begin to disintegrate, the new colonies link to form the structure of the butterfly's body.

At some point, an integrated supercolony of transformed cells that had once crawled along the ground emerges from the cocoon and flies off into the spring sky on multicolored wings. If there is a positive image of the future of human-computer relations, perhaps it is to be seen reflected in the shapes of the imaginal cells of the information culture—from eight-year-olds with fantasy amplifiers to knowledge engineers, from Ted Nelson to Murray Turoff, from Clyde Ghost Monster to Sourcevoid Dave, from ARPA to ORIGINS.

The flights of the infonauts are not the end of the journey begun by the patriarchs, but the beginning of the most dramatic software odyssey of them all. It is up to us to decide whether or not computers will be our masters, our servants, or our partners.

It is up to us to decide what *human* means, and exactly how it is different from *machine*, and what tasks ought and ought not be trusted to either species of symbol-processing system. But some decisions must be made soon, while the technology is still young. And the deciding must be shared by as many citizens as possible, not just the experts. In that sense, the most important factor in whether we will all see the dawn of a humane, sustainable world in the twenty-first century will be how we deal with these machines a few thought up and a lot of us will soon be using.

I'm working barefoot today, under the plum tree in my garden. My toes touch the lawn while my mind navigates cyberspace. I type these words on a laptop computer a thousand times more powerful than the suitcase-sized contraption I used to write *Tools for Thought* in 1983. I need to list a few contemporary technologies that were unknown in 1983, so I ask a virtual community on the Internet for help. I look up at the ceiling, and see the sky.

When I first logged on to a computer bulletin board in 1983, the box that connected my computer and my telephone line cost $500. Today, a device thirty times faster is a quarter the size of that first modem and costs less than $100; cable modem speeds hundreds of thousands of times faster are now available to millions of households that use coaxial cable to receive television programming. Over the last twenty years, the world has been webbed and interpenetrated by copper, cellular, packet radio, coaxial, satellite, submarine, microwave, and fiberoptic media pipelines.

There is still a serious "digital divide" between those whose circumstances provide access to the world's wiring and those who can't afford a telephone. Half the people in the world still haven't made a telephone call. At the same time that so many people live in the pre-electronic era, more people than ever are leading technology-augmented lives. The "digital divide" and a "knowledge society" are both emerging at the same time.

A few minutes after I posed my question about 1983 technology to my virtual community, I checked it again and saw that people in other parts of the world had added cell phones, home fax machines, ATMs, and civilian global positioning satellite receivers to the list of today's tools that didn't exist then. Early 80s videogames were two-dimensional, low-resolution shootouts with aliens, such as *Space Invaders*, mostly played in arcades; today, *Quake* is a high-res, photorealistically gory shootout with other players, linked in real time across the network.

In 1983, you had to hire a private detective to find someone, or to gather extensive personal information about someone; today, if you know how to search the Web, you can do your own investigation. We feared Big Brother when the mythical date 1984 loomed, but never suspected the emerging power of Little Brother—citizens who use the Internet to snoop on each other.

A world-wide web of interconnected multimedia networks and hundreds of millions of powerful personal computers have turned the visions of the future I described nearly twenty years ago into the reality we inhabit today. New industries, new ways of life, new ways of working and thinking, new engines for creating wealth, have emerged. The people I wrote about in 1983 created a new kind of world. We're only beginning to learn how to live in it.

The world today differs in several major respects from the world the visionaries foresaw. I can testify that my life in particular has been transformed for the better by the personal computer and the Net. I do have moral reservations now that have grown since I first wrote about computers as mind amplifiers, but those reservations aren't focused on the ways my own work and life have been enriched. I can sit here, wriggling my toes in the lawn, while I edit drafts, communicate with other minds, collect data from every place on the globe. "What about the rest of the world?" is still the most important question, as Licklider and Taylor warned in 1968, on the eve of the launch of the ARPAnet.

I'm not sure whether the people who inhabit cubicles in skyscrapers, stare at screens most of the day, and process richer people's credit card transactions feel as liberated as I do. I'm certain that the teenage girl assembling computer hardware in a maquiladora in Ciudad Juarez is even further away from my experience of increasing personal freedom. But my own life and career have changed in beneficial ways since I got my first bank loan—ten thousand dollars to buy an IBM XT PC and impact printer that sounded like a machine gun—in 1983.

The tools I've learned to use over the past two decades have enabled me to think, communicate, learn, and earn a living in ways I never dreamed of back in the final days of the typewriter era. In 1983, I sought out the people who invented mind-amplifiers and asked them about the future. By now, I've been using their inventions for most of my career. My way of thinking in the middle of my life has been shaped by the computer and network, just as my mind was shaped in my youth by the book and the library. My youth was spent in the final decades of the Gutenberg era, my middle age was shaped by the first decades of the computer revolution. Will coming decades bring equally radical changes in the way we think and communicate?

The occasion of the MIT Press's new edition of *Tools for Thought* affords an opportunity to look at my own past predictions as a journalist of technological change, recall the dreams of the visionaries I profiled back then, and compare our versions of how the future would look with the way the future actually turned out. Retrospective futurism is so much easier than the prospective kind.

Some predictions of the first paragraph and other pages of *Tools for Thought* have borne out remarkably well. The world certainly has changed dramatically because of personal computers and the Internet, as Engelbart, Nelson, Licklider, Taylor, and Kay predicted and worked to make happen.

The original visions of augmented knowledge communities have still not been fully realized. The most important changes in the way we think, communicate, work, learn, and live together because of our use of these new media are still not deeply or widely understood. And my own perspective on technology has grown increasingly critical as the personal computer, Internet, and anywhere-anytime-any-medium communications revolutions have triggered avalanches of social changes—not all of which turned out to be beneficial or benign.

Several of my predictions and those of others were wrong or were off in some way, large or small. I'm interested in what we can learn from those forecasting errors. In retrospect, the biggest flaw in my thinking in 1983 was the assumption that nearly everyone shared at the time, and which is still the commonly accepted wisdom in much of the world—that technological progress, especially in communication media, was not just inevitable, but would be an unalloyed positive social benefit. More people these days are beginning to perceive the ways in which the bright light of progress casts shadows on our freedom, our privacy, our human relationships.

The drawbacks of a global network of mind-amplifiers are more easily visible now that we are living in one rather than imagining what it would be like. What can we learn from yesterday's dreams that will help us live in today's world, and refine our vision of tomorrow?

Doug Engelbart, bless his stubborn perseverance, is still actively pursuing his dream of intellectual augmentation, a notion that seized him as a young engineer in the 1950s and has kept him focused on the task of making it real for half a century. He's no longer the lonely, uncelebrated prophet. His "Bootstrap Institute" (http:/./www.bootstrap.org) is trying to "raise the collective IQ of organizations." Because Engelbart's vision was the pivot of the story I set out to tell in Tools for Thought, in the summer of 1999 I traveled to the present offices of the Bootstrap Institute, which are generously donated by Logitech, a mouse manufacturer on the out-skirts of Silicon Valley. I talked face to face with Engelbart about where we've come over years past and where we still need to go in years to come. I also communicated via telephone and email with four of the other people I profiled in 1983—Avron Barr, Alan Kay, Brenda Laurel, and Bob Taylor—to see what they think of the changes in the world they envisioned and tried to construct.

Two predictions that erred on the optimistic side were about the role of personal computers in public education and knowledge engineering as a very large commercial enterprise. In 1983, I was among those who thought that computers in schools could revolutionize education. Now, decades after the failure of personal computing to make a dent in the problems of the public education system in the United States, many people who don't seem to remember that episode believe that Internet in schools will revolutionize education. Having watched one of my fondestly hoped-for forecasts fail, I'm now more capable of seeing the gaps in the reasoning behind those yet-unfulfilled, and perhaps unfullfillable, hopes for information technology in education.

Even if sufficient resources could be allocated to making the best educational use of the Net, the problem remains of teaching tomorrow's citizens how to think for themselves. Parents aren't keen on teaching their children a critical and questioning attitude, and teachers aren't equipped to show children how to find their own way through the new world where you don't just trust the publisher of information to tell the truth. It seems, even to an optimist like myself, unlikely that the political structure of the public school system ever will allow the kind of training and critical thinking necessary for the most effective use of computer networks in education. Perhaps the real technology revolution in learning will come through the kind of "subversive media" Alan Kay is still designing.

Knowledge engineering was never the gigantic industry its proponents predicted. The expert systems entrepreneurs and I were overconfident about the pace of evolution of sophisticated software for capturing and making human expertise widely available. In 1983, "knowledge engineer-ing" referred to the process of transferring knowledge from an expert to a computer program, which would then make that expert's judgment about specific subject domains available to non-experts. That vision never really came about on a wide scale, although expensive specialized expert systems are in use today.

Avron Barr has transformed his knowledge about computer programs that could substitute for human experts into ways of building computer programs that amplify the capabilities of groups. I talked with Avron Barr again, about the ways in which knowledge engineering and his own thinking have evolved in recent years.

Atari Laboratories and Atari, Inc., beneficiaries of the first videogame craze, failed in the personal computer boom of the 1980s, and are now footnotes to the history of the PC Industry. MIT's "Architecture Machine Group," where Atari Labs' alumni Scott Fisher, Steve Gano, Eric Hulteen, and Mike Naimark had studied in the late 1970s, grew into the Media Lab of the 1990s, led by Nicholas Negroponte. If PARC was the place where people "invented the future" in the 1970s, Media Lab might have played that role in the 1990s. It looks as if the equivalent for the first decade of the 21st century has to be the one company that invests the most money in fundamental research, and which most aggressively works to avoid the same kind of mistake Xerox made — Microsoft Research.

Atari Labs' director Alan Kay was a fellow at the Media Lab, along with Marvin Minsky and Seymour Papert. Media Lab researchers have taken advantage of advances in miniaturization to embed microprocessors into building blocks for its "Lego Logo" research program. Alan Kay envisioned an imagination amplifier that schoolchildren could use, but not even he fully foresaw the advent of embedded intelligence and ubiquitous computing, in which children's toys and household appliances each contain more computing power than the entire Defense Department had in 1955. It might not be a surprise to those who knew his early work to learn that Alan Kay is working for Disney now.

Brenda Laurel, leader of the "Future Squad" at Atari Labs, joined another 1990s version of PARC, "Interval Research," then founded the company that created the first line of products to emerge from that research — software specifically designed for pre-adolescent girls. I talked with Brenda about the way the world looks after an additional decade and a half of research and entrepreneuring beyond the Future Squad days.

The Perseverance of a Long Distance Thinker

It all started with Doug Engelbart, who looked a half century ahead of his time and saw what nobody else could see — a world of networked mind amplifiers, decades hence. When Doug Engelbart first started writing about "the augmentation of human intellect," Silicon Valley was still home of the world's richest fruit orchards. The first time I met Engelbart, it was in a modest office building belonging to TymShare, an early online data services company. TymShare had bought Engelbart's Augment system and hired Engelbart after his research sponsors stopped supporting the Augmentation Research Center. TymShare occupied a modest building, overshadowed by the larger campus of Apple Computer.

Engelbart was soft-spoken, hopeful, inspired, inspiring, determined, and resigned to the fact that he was almost totally unknown to a population that thought Steve Jobs invented the personal computer. Money and fame never were important to him. What really mattered was to help people collectively solve complex problems. By the 1990s, however, as the world came to resemble more and more the vision that he first experienced in 1951, Engelbart was discovered by the media and recognized by the world outside the exclusive fraternity of ARPA veterans.

On December 9, 1998, the Memorial Auditorium at Stanford University was packed by a standing room only crowd, assembled for an extraordinary one-day symposium on "Engelbart's Unfinished Revolution" (http://unrev.stanford.edu/). The old-timers were there, those who remembered the fabled 1968 demonstration at the Fall Joint Computer show. Some of them were now running the most powerful technology companies in the world. The superstars of the new world of the Web were there, including Marc Andreesen, creator of the first popular Web browser, "Mosaic," and co-founder of Netscape. The day was filled with panels and speeches, and a screening of the '68 demo, but the emotional highlight was when Engelbart stood alone on the stage to receive a prolonged standing ovation, his eyes glistening with tears.

Around the time I first interviewed Engelbart, Tymshare was acquired by aircraft manufacturer McDonnell-Douglas. Despite the fact that yet another parent company did not understand the longer term value of Augment and the system that Engelbart was still working to build, Engelbart, characteristically, remembered that era in terms of what he learned. "It was very good experience for us," he recalled when I interviewed him in 1999,

because this enabled me to start interacting with the people doing design, manufacturing, and support for aircraft. The underlying question for me always has been how can we get collectively better at handling complex urgent problems? Working with computer-aided design in an aircraft manufacturing company, I saw that the standards for knowledge containers have to help the engineers talk with the manufacturing guys, help the manufacturing guys talk with the suppliers—

who also have engineers and salesmen and planners, and with the contracts people who have to negotiate and support contracts. That's where the idea of an open hyper-document system emerged.

By the late '80's, it became just so clear to me that we'd have to get cooperation from a number of organizations to build the system I envisioned, even if they were companies that competed with each other. That's when we started the Bootstrap Institute. We set up an independent flag, and said, "Here's a strategic approach that collections of organizations—including those that usually compete—could get together and go after."

The CODIAK concept of how to build a dynamic knowledge repository come out of that work. The acronym derives from the expression "COncurrently Developing, Integrating, and Applying Knowledge."

It's one way to chunk together the set of capabilities the future organizations are going to have to be a lot better at. We also started using the term "collective IQ" to apply to the cognitive and communication and organizational changes that had to be built. You can have all kinds of computer-aided design, and planning software and communications, and everything else in there, but the integration aspect—how it all works in the daily lives of the people who use it, and in their organizations—is harder to pursue than technical goals. But it's essential. It's the biggest part of what we have yet do. We need to stimulate active coevolution of the human system and the tool system.

We started out being emphatically explicit about the difference between the tool system and the human system. The human system embodied all the practices and conventions and vocabulary and skills and roles that were enabled by the tool system. We did note that the co-evolution of the human and tool systems is the driving force of the kind of intellectual bootstrapping we set out to do. When people adopt powerful new tools. a subsequent co-evolution of the human systems is inevitable. Very seldom do you come in and just automate something and find that your change has had no other effect. The goal of augmenting rather than automating is being able to stimulate and design this co-evolution.

Society needs to focus on improving improvement. An organization's core business is what we call the "A" activity. There are activities that can improve A work. What we call the "B" activities are the ones taken to improve the way an organization does its business—improvement activities. But in times of rapid change in the ways work is done, you've got to consider how you improve that improvement— the "C" category of activity for any organization. Whatever else you do, there has to be the conscious activity of improving the improvement process. We discovered that is a key concept in strategic thinking about augmentation.

I asked Engelbart to comment on the way the world has adopted the GUI, the graphical user interface that was one of the more spectacular developments to emerge from his pioneering work with point-and-click interfaces. "I don't like it!" Engelbart exclaimed, the second I said the word "GUI."

> I find people sort of connect me with that, but when we first started building our systems, we arrived early at the design question "Do we use menus or what?" By then I had enough experience with the chording keyset to just take a look and say, "In the time it takes me to move my cursor up to click on the menu item, I could hit two or three characters with my left hand on the keyset while pointing with my right hand on the mouse." In fact, if you point at things with one hand and say what you want to do with the other hand, you can work far more efficiently than today's GUIs allow. The Augment system let individual users develop a big repertoire of verbs and nouns. It started to be a command language, and when you look at the GUI as it is today, I just think it's like a very primitive pidgin English. It's like grunting and pointing instead of speaking.
>
> One of the design principles we adopted in the late '60's was to provide a range of user interfaces, depending upon the kind of skill that a user has. We created a profile of verbs and nouns that could be set for you. If you're a beginner, you get a very simple subset. Then, as you want to learn more and more, you can add more. This was part of our picture of what an open hyperdocument system would be. If people wanted to use today's GUI, sure, fine. If they want to use a more powerful vocabulary and set of processes, they're free to do that.

"What do you think of the way the Web has developed?" I asked.

"It's a start," he replied. "It's important that we've found some sort of combination of tools that is simple enough and usable enough that it's caught on the way the Web has. But what is also very, very important is that we don't get arrested at this level of development. How do you evolve the Web into something that could be really effective?"

I asked him to name some essential principles that he articulated, but which the Web has yet to adopt.

> Integrating the editor and the browser is one. It's ludicrous to do your authoring in one kind of environment then go to an entirely different environment to browse. Another principle is that every object needs to

be addressable intrinsically, so you don't have to go insert a special tag. You ought to be able to link to a word or sentence or paragraph or graphic. Another principle is the option of having different ways users can view a document, like seeing only the first line of every paragraph.

Some of Engelbart's former colleagues are now top managers at companies like Sun Microsystems and Netscape. The Bootstrap Institute started to build a consortium of companies to contribute money and people and a commitment to experiment with the collective process of learning how to improve improvements—the "Bootstrap Alliance."

Doug Engelbart is still soft-spoken and self-deprecating, although many who have worked with him can attest to his stubborn conviction that the overall scope and potential of the digital revolution is far greater than is commonly recognized, and that the rest of the world should be energetically going after the "collective IQ" he seeks. "That would let me retire, if I could see a really effective pursuit of that collective IQ-raising." I mentioned to him that I couldn't think of one person who has pursued one goal for 50 years, the way he has. Engelbart thought for a moment, then smiled and replied softly: "They must think 'Geez, that guy must not have much imagination, he can't think of anything else to do!'"

Looking Back on the New Old Boys

Bob Taylor, now retired in the San Francisco Bay Area, together with the late J. C. R. Licklider, wrote presciently about the future of networked computers as early as 1967; and in 1966, he initiated the ARPAnet project when Taylor was an office director for the Department of Defense–sponsored Advanced Projects Research Agency. Later, at Xerox Palo Alto Research Center (PARC), Taylor built and led the team that created an astonishing number of firsts: the graphical user interface, point and click word processor, Ethernet local area network, laser printer, desktop publishing, and internet. PARC's work was enabled by the prior work of SRI's Engelbart and ARPA, but they assembled the components of what has become today's global computer infrastructure—visual computers, linked via an internet—into a commercially viable system.

Recent books and articles have since recognized that Bill Gates took the idea that turned into Windows from PARC, just as Apple's founder Steve Jobs took from PARC the idea of a graphical user interface that turned into the Macintosh. But PARC's role in creation of the Internet is less widely recognized. In late 1993, just before the publication of *Tools for Thought*, Taylor and more than a dozen key researchers from his PARC team moved out of Xerox entirely and built the "Systems Research

Center" (SRC) at Digital Equipment Company (DEC). The DEC team beat the first truly powerful web search engine, AltaVista. They created a networking scheme for real-time video and audio over a global network and created Modula 3, a language that became the precursor to the Java language.

Taylor retired in 1996 and still stays in touch with much of the remarkable research community he led for thirty years.

From Taylor's "Computer Systems Laboratory" at Xerox PARC came Bob Metcalfe, the founder of 3COM. John Warnock and Chuck Geshke founded Adobe, the computer typography and page design company that created the desktop publishing industry. Charles Simonyi, who took the "Bravo" word processing software he created for the PARC Alto and transformed it into Microsoft's "Word" product, the most widely used word processor in the world, is a billionaire today, or close to it. Alan Kay is a Disney Fellow. "Some of my former colleagues work SRC, some at Microsoft, some at Adobe, and some at Sun Microsystems. One, Jim Morris, is dean of school of computer science at CMU. There's a great community out there. I am proud to be a part of it," said Taylor, when I re-interviewed him in 1999.

When I asked Taylor if history has overlooked anything significant, he responded with a provocative assertion:

> The first internet was also built at PARC. We connected PARC's Ethernet and the ARPAnet in 1975. That's what an internet is—two or more separate computer networks connected together. Vint Cerf and several others claim to be the father of the Internet. Cerf pulled together a committee at Stanford around 1975 to create what eventually became TCP/IP, the fundamental protocols that enable computer networks to communicate over the Internet. He invited some people from PARC— Bob Metcalfe (who was largely responsible for the creation of Ethernet, and later founded 3COM), John Schoch (now a Silicon Valley venture capitalist), David Boggs, and others—to help the Stanford committee with the design. But PARC lawyers told our people that they couldn't tell the people at Stanford what we had been doing with a working internet.

According to the account Taylor gave me, the PARC researchers who had already solved the problem of internetworking met with the Stanford group who were trying to come up with a fundamental software architecture for connecting networks together. According to Taylor's recollection,

The Xerox guys said things like "if you implement that design, what happens if this other thing occurs?" The Stanford guys would say, "I guess you're right. We need to rethink that design." They went through this scenario several times, with the PARC researchers questioning ideas that the Stanford team originated—pointing them toward what might be wrong with their proposed technical solutions—but because of their legal restraints, not laying out what they knew about exactly how to build an Internet. One of the Stanford guys finally said: "You guys have already done this." And indeed, Ed Taft, now at Adobe, had already completed PUP (PARC Universal Packet) at PARC before TCP/IP was created.

When I told him about Doug Engelbart's and Alan Kay's critical remarks about the design of the World Wide Web, Taylor said, "I agree that the Web is a bad implementation, but it's also a miraculous piece of work. The Web protocols make a bunch of software in different systems around the world work together. The Web protocols didn't create a great meal, but it did make an edible meal out of garbage. Yes there is a lot wrong with it, but it's remarkable nevertheless. In order for there to be an Internet, you need a PC and a local and a global network, and it helps if you understand desktop publishing and graphical interfaces. We had created and put all those ingredients together by the mid to late seventies. But the Internet didn't take off until mid mid-'90s."

I asked Taylor why it took so long for the technology to spread through the population: "A lot of people screwed up along the way, especially Xerox," he replied. "Xerox had the opportunity to put reasonably inexpensive networked PCs into people's homes years before Apple or Microsoft existed. Apple took advantage of Xerox's slip, but Apple blew networking. Xerox made some mistakes, Apple made some mistakes. We had all the pieces but nobody could put them together until 20–30 years later. So my own predictions were way off. I thought that by the end of the '70s that the world would have the computing and networking infrastructure we didn't end up having until about 1993. I missed by 20 years or more."

What did the original designers do right?

Many people today are worried about the Internet becoming less democratic, perhaps centralized or under a few people's control, at some point in the future. I don't think that's going to happen because of Wes Clarke, who came up with the idea of the IMP, the Interface Message Processors that enabled the packets to find their way around the network without central control. Larry Roberts and other people were talking about having ARPAnet controlled by a central computer in

Nebraska. IMP makes it very difficult to centralize control of the network. If there had been one computer controlling ARPAnet, we would have already had cause to worry. That single architectural foundation to packet-switched networks makes it difficult for anyone to control the Internet. Nobody ever calls Wes Clarke the father of the Internet, but he has as much right to that title as anyone who claims the title today.

What new technological revolution is likely to change the world in the foreseeable future? Taylor agreed with me that it was far easier to foresee technological futures in 1983 than it is today, but added: "I can see a continuing convergence of Internet, television, telephone communications, including mobile telephones, and even appliances. When every little gadget imaginable has a computer in it, and all those embedded computers are talking to all the others—that's going to make another huge leap possible."

The failure of the Xerox Corporation to capitalize on the ten-year lead that PARC gave them in PCs and internetworking has been documented in books such as *Fumbling the Future,* and the history of PARC's largely unrecognized role in creating the basis for today's high-tech industries was covered in a book, *Dealers in Lightning.* Taylor points out that PARC research still contributed enormously to the parent company's bottom line: "Xerox got a billion dollar business out of laser printers alone, which paid for all the research costs at PARC, and much more."

Taylor doesn't see a lot of growth in the kind of truly fundamental research that PARC pioneered. "Microsoft is doing PARC-like stuff, and IBM still has its research entity, but there isn't a great deal of PARC-like research in major companies around the world. Chuck Thacker, Butler Lampson, Mark Brown are at Microsoft research. Microsoft really does have some very large investments in research, not just advanced development, which is what many companies call research these days." In other words, the company with the largest investment in "creating the future," PARC-style, appears to be Microsoft.

Toys That Think and Other Subversive Media

Alan Kay still believes in powerful ideas, still directs his energies at creating a new medium for stretching young minds, and still works with a small crew of wizards who "deal lightning with both hands." Today, he continues to design "subversive media," now under the aegis of Disney, just as he and a small group of remarkable minds in the employ of Xerox invented the networked personal computer thirty years ago.

Kay was running Atari Research Lab at the time I first met him in

1983. He was Atari's Chief Scientist and Vice President of Research and Development: "Atari Labs was my only attempt at running a large organization," Kay recalls. "I learned quickly that I didn't want to do that. At Apple and Disney, the group has been roughly size and scope of the group at PARC in the 1970s. I'm doing the same thing today I started to do in the 1960s—trying to discover ways to help children grow up to think better than most adults do today."

The Dynabook was the dream that Kay brought to Bob Taylor's Computer Systems Laboratory at PARC in the early 1970s, a vision of a handheld device with high-resolution screen, wireless network capabilities, and sophisticated graphic software, a vision that drove the development of the first point-and-click interface—the personal computer as it is known today. Even after Xerox "fumbled the future" by failing to market PARC's inventions, it has taken decades for the global computer and communications industries to approach the level of sophistication of the devices Kay and colleagues specified a decade before the first Macintosh came off the production line.

"Our specs for the Dynabook were a little less ambitious than today's technologies turned out to be, in some respects, and more ambitious in terms of others, especially the software," Kay recalled, in our 1999 interview. "The original Dynabook idea was 95% software. The other 5% was the lightweight package for delivering the software. The state of the art of software for children of all ages that embodies truly powerful ideas is still not there—not where we envisioned it would have to be. The closest thing I know of today to what we saw in 1970s is what we are doing at Disney." Kay couldn't talk in specifics about what the Disney group, including Danny Hillis, pioneer of distributed supercomputing, were making. But he did say that the products were only months, rather than years, away from public view.

Kay tends to talk about what "we" thought about personal computers when "we" first started thinking about them, referring to the ARPA projects in the late 1960s (especially the creation of ARPAnet) and the Alto, Smalltalk, and Ethernet projects at PARC in the 1970s, when the world's entire braintrust in networked personal computing worked together on outlandishly futuristic inventions that ended up changing civilization.

> We were interested in what there is about computers that no other medium can do. What capabilities of this new medium would enable people to think in an altogether different way, not possible with previous media?
>
> We thought that just as the press enabled people to argue about things

beyond the scope of traditional oral discourse and manuscript transmission, a dynamic book would allow people to argue about science and politics and other things in ways the printed book didn't make possible. It's not possible to argue for democracy in terms of stained glass windows and hand-written documents, but you can argue for theology and monarchy very well with those pre-printing media. The printing press and authors such as Tom Paine made it possible to make issues of human rights important politically. It was no accident that the press and science preceded our modern form of democracy.

Kay argued that the particular way of thinking necessary for public discourse and self government was apparent to scientists because the enterprise of science is built upon the freedom and ability to argue.

A literate public is not an oral culture plus a writing system. The epistemological stance that goes with literacy is different from the way of thinking about knowledge that goes with orality. The act of learning to read, to deal with abstract symbols, leads to a different way of thinking about things—a way of thinking that is necessary (but not sufficient) for science and democracy.

This idea of a new kind of literacy and way of thinking that could be made possible by a technology that didn't yet exist was extraordinarily important to me when I realized it. I realized a lot of it from seeing Seymour Papert's stuff in '68. The goal of Dynabook was to guess what this new literacy would be like and make a form of it that children could learn, and put it into a children's computer. It would have to be portable because kids are mobile. The individual units would have to be in communication. Wireless networking was built into the original specs. Kids want to design and create, so we wanted bitmapped screens and ways to create graphics. Because we didn't want to throw out the book and print, we wanted high resolution screens to do readable fonts, a capability we forecast would become possible when PC displays could put a million pixels on a screen. The typical large screen on a portable display today is around 800,000. Within the next 5 years, affordable PCs will become as effective visually as print on paper.

Kay continued, in our 1999 conversation:

In the late 1960s Nicholas Negroponte and I were in a very small minority who thought the media aspects of the computer were really important. Again looking at the history of the Gutenberg press—the ability to print things was not important as the rhetoric invented to take advantage of what press could do—the argumentative essay and scientific essay, 150 years after printing technology emerged. Would we have to wait for a century and a half for people to bumble into the power of computers as media, or could we take a lesson from history and invent the media of the future?

The greatest product of the post-printing way of thinking thing accomplished in the 17th century was Newton's Principia. The Federalist papers in the 18th century were another world-changing application of the new rhetoric that widespread literacy made possible. A powerful set of mental tools was developed when people applied formal systems that had to do with numbers and space to written argument. Science and democracy were among the things that could only be built with these thinking tools.

Along with understanding the computer as a medium is an understanding that the content of computer is language. Bit patterns are unique symbols that can stand for many different ideas, depending on how they are interpreted. It had occurred to Von Neumann in the 1940s that computers would be capable of simulating the physical universe in ways that not even classical mathematics could pragmatically do. Back then, they were simulating artillery trajectories, but he could see astronomy calculations turning into a language machine of a mathematical form based on differential equations. Other complex non-linear phenomena could be represented by computers in ways that could lead to new understanding.

Kay acknowledges that it has been a difficult problem, translating Papert's insights about "powerful ideas" and his own understanding of computers as language machines into something that could improve the minds of millions of children: He added that revolutionizing traditional schooling is not the only means of affecting millions of minds. From the beginning of the Dynabook quest, Kay has thought in terms of "subversive media" that bypass established institutions.

The printing press was subversive because people discovered quickly that you could print many things, not just bibles. And you could read a book by yourself and not in a dogmatic social setting. Over the years, we have made a lot of different forays into simulation media. The Vivarium project at the Open School in Los Angeles in the mid-1980s lasted seven or eight years. Papert has continued to experiment, and so have others at the Media Lab. The personal computer and early Internet industries were selling to customers who didn't fully understand what they were buying, which made them something like a teenager just beginning to get interested in music—unsophisticated taste, combined with real buying power. The result was the entrenchment of poor de facto standards. MS DOS wasn't the right way to present computation power to new users, but it became a standard.

The World Wide Web is kind of the MS DOS of the Internet. Engelbart had thought out what an open hyperdocument system should work like. The creators of the Web did not understand what Engelbart had done, and apparently lacked the intellectual curiosity to find out. But the saving grace of the systems that dominate the market is that none of them prevent users from doing their own thing. You can invoke a program from MS DOS or the Web that has nothing to do with their flawed design standards. People have remarked about television that it doesn't tell people how to think, but because it is so total, it makes it difficult to think about things other than what it is presenting. People who are not strong thinkers find it hard to think beyond MS DOS or the Web. Most people don't know that Net and Web are two completely different things at two different levels. The Web is just one example of the kind of thing you can build on the Net.

One approach to education and technology that succeeded was Frank Oppenheimer's Exploratorium in San Francisco, because he designed it as a playpen instead of a museum. The Exploratorium is an example of subversive media that transform education without being part of the schooling system. Over the years, Seymour and I and other people have concentrated at times on schools and there have been times when we concentrated on subversive media. This is a good time for subversive media. That's why I am at a media company.

Like Taylor, Kay is surprised at how long it took for PARC innovations to spread: "I published 'A Personal Computer for Children of All Ages' in 1972. From my standpoint, things are ten or more years behind where they could have been."

The Future Squad's Future
Brenda Laurel's group at Atari Research, under Alan Kay's direction, explored future directions in human computer interface design—imagining what could be done with a wall that can hear and respond to commands, a computer that knows where you're looking, head-mounted displays that immerse you in a three-dimensional virtual world. After Atari Labs closed shop, Scott Fisher went to NASA, where he pioneered virtual reality (VR) research. Eric Hulteen went to Apple's Human Interface Group. Laurel organized the publication *The Art of Human-Computer Interface Design* (Addison-Wesley, 1990). Then she participated in the "Guides" project at Apple, an attempt to include storytelling in the human interface—a concept that still has not been fully explored, using

onscreen characters to provide narrative guidance through, for example, a historical simulation.

In 1989, Scott Fisher and Brenda Laurel teamed up to create a VR company, "Telepresence Research," funded by Japanese investors. Although Scott Fisher continued with the company, Laurel eventually left to pursue further research in the use of VR beyond the world of simulation and training. In 1991, she wrote *Computers as Theatre*, an influential book about interactive theory and design that is still required reading in university computer science and new media departments. In 1999, she and I sat under the plum tree in my garden—a tree she gave me as a sapling, not long after the publication of *Tools for Thought*. I asked her to recall her journey over the past decade:

> I went into VR research in order to show the potential of the medium to do something besides vehicle and combat simulation. I wanted to create something that would change the way people thought about what the medium could do.

Laurel and Rachel wrote a proposal for a virtual environment that would "treat the relationships between landscape and narrative as a kind of playground," and submitted it to the Banff Center for the Arts. By the time the proposal was accepted, Laurel and Strickland had started working for a new laboratory called Interval Research.

Billionaire Paul Allen, Bill Gates's original partner, started out in the mid-1990s to create the next PARC. The founding myth of Interval Research was $100 million in funding for ten years, and a sponsor who would avoid the mistakes of SRI and Xerox. David Liddle, one of the original PARC golden boys, was the leader. Many of the Atari Labs Future Squadders ended up at Interval Research: Mike Naimark, Rachel Strickland, Eric Hulteen.

A couple years after Interval was founded, though, college kids started building websites and becoming billionaires. According to Laurel, the atmosphere at Interval changed over the years.

> I think that the fast bucks being made in the Valley tempted Interval to start spinning off companies earlier than it was really ready to. The infrastructure and mentality to start new businesses was not in place yet, and a huge cultural transformation needed to occur inside Interval to move projects successfully from research to business. Despite some very talented people, I don't think Interval ever got good at making that transition.

I was able to bring in a big piece of funding from Interval to the Banff project. I was lucky enough to be working with Rachel Strickland and Mike Naimark, so all three of us were colleagues again after the Atari Lab. What we conceived had a lot to do with the work that Mike had been continuing to do with movie maps. [Naimark had been instrumental in making the first movie map, the Aspen videodisk project that MIT's Architecture Machine Group had done in the late 1980s.]

Rachel brought a theory about the relationship between landscape and story, place and story. I brought a notion of play and a notion of drama in natural settings, and I think we shared a sense of the mythic that energized the work. We found three natural locations in the Banff area that were sacred to the people who lived there, indigenous people who had stories about these places. Mike designed different techniques for capturing and representing each of those environments.

Then within these environments we placed four archetypal characters that seemed to have similar meanings in different cultures. For example, a fish is often the bringer of news. The spider is often seen as a creator. The snake has secrets. And Crow, or Raven, has a bird's-eye view. He can see in the large. He's also attracted to small shiny things, so he can zoom in to notice details.

We put those characters in the world, and the kind of fun innovation that I think I contributed to that was the idea that these characters could become smart costumes that you could slip into through the virtual environment, take on the character's body, take on the character's voice, be seen by other people in the world as that character, and move around as that character.

But fundamental enabling technologies for making VR economically viable—cheap computation, high-resolution displays—would take years to develop. Laurel turned her Interval research to a question that had been close to her heart since she started in the computer gaming industry in the early days of PCs—the question about why so few young girls use entertainment software, although adolescent boys support billion-dollar companies, from Atari to Nintendo. Finding out the answer to that question could lead to a very large business opportunity.

We partnered with a market research firm, Cheskin Research, who helped us explore the dynamics of play. The research happened in three stages. First, we did a huge literature survey of all of the domains that we thought might shed some light on the problem. So we looked at play studies, gender studies, neural physiology, everything that we could find out about sex difference and gender difference as it relates to play that had already been discovered by somebody. We made a giant database of that and looked at all that material very closely. We selected the most promising ideas to follow up.

In the second stage we interviewed the experts who were responsible for creating that interesting data, and in some cases some other adult experts who ought to know something, like the head of Barbie marketing at Mattel, or people who designed other products for the demographic. In the third stage, which was by far the largest part of the research, we interviewed around 1,100 kids all over the country, both boys and girls, about how play differs by age and gender in this country, what girls and boys think about technology and technology play and where the differences and similarities are, how boys and girls know that products are for them. We did this in several waves, so that by the time we were near the end of the interview section, which lasted about two years, we were looking exclusively at girls, asking very pointed questions about computer games, and showing them at first paper prototypes of things that they could play with, and then later, software prototypes. We narrowed our focus down to actually forming some hypotheses, building examples of them and testing them with kids.

After three years in, we translated our findings into design principles. If you're going to build a game for a girl, here are some design principles that you probably ought to follow. And right in that last stage was when the decision was made, to spin off a company.

Something unforeseen happened between the founding of Interval and that decision to make software for girls—the explosion of the Web business. Netscape and Yahoo were coming from nowhere and growing to billion-dollar companies in months. According to Laurel, "Interval was curiously immune to the Web. Looking back if we had launched Purple Moon as a Web company it would have been a very different scenario."

Purple Moon, the girl games company Laurel founded, backed by Paul Allen, focused on the CD-ROM business, which turned out to be a disastrous mistake because that business was in the process of tanking. Purple Moon launched with two products that did well for going up against big brands in a brand new market, but Mattel launched the Barbie product a year earlier, and Purple Moon was beaten to market by a year by a very large company with a forty-year-old brand franchise. Purple Moon launched a Webster at the same time, which was spectacularly successful, the third-largest children's site on the Internet, but wasn't generating revenue. In order to keep the company going, they started accruing debt.

We wildly launched into outer space without a good enough idea of how we were going to ground the business and how we were going to move it forward, and I think there was some denial going on about market conditions. The Board was fixated on an Internet-style valuation and wouldn't take investment at a more realistic level. Internet companies can base gigantic valuations on ideas like "market cap," but companies that actually make things are usually valued as a multiple of annual revenues. Early on, there was interest and respectable valuation from some big companies, but the Board held out for the big bucks. They preferred to borrow money rather than take investment at a lower valuation. But that was eating the future—who wants to acquire a company with a big debt? We continually under-invested in our Webster in order to make more and more CD-ROMs that were less and less profitable. So the web business didn't grow fast enough for the company to make a graceful transition. In the end, I feel that Paul Allen pulled out short of his commitment. When his folks finally figured out that they weren't going to make a quick killing, they just slammed us against the wall. We had just shipped our eighth title and launched e-commerce. Mattel was sniffing around again to invest. We had to fire 45 people and lock the door, just like that. To be fair, I think I probably contributed to the mess because I was so opposed to the idea of getting in bed with Mattel early on, when they first offered to invest. I had such cultural issues with the Barbie flavor of what it means to be a girl and what girls want that I was resisting doing business with Mattel on principle.

Ironically, by the time the company found itself sinking without a lifeboat, Mattel stepped in to buy Purple Moon at bargain-basement prices.

I asked her about the future:

The good news is that a lot of enabling technologies are coming along wonderfully, like speech and movement recognition and 60 frames a second VR, and eliminating lag in VR systems, and latency problems going away. We have cheap, fat bandwidth and the ability to download enormous amounts of information into storage devices that you and I can finally afford. All these technological developments hold great promise. It's the business and social issues that are retarding our ability to build compelling examples that will influence what we do with those technologies. The idealism that we all had 20 years ago isn't enough to make that happen. That's the bad news. There is no Camelot where we can go build the compelling example now, not without having a business plan that shows how we can get to an IPO in six months.

She was more pessimistic about the Web as an educational technology:

> It isn't about technology anymore—it's about culture. The Web is only as good as the culture it's part of, and the value of it to a child in a school is only as robust as the material made available there, or the interactions that can happen there, that come out of the hearts and minds of people, and out of companies creating stuff. A culture that can't find its values with both hands isn't going to know how to make use of a tool like the Internet for a child in an educational environment. We're living in a society that doesn't value its children and doesn't value thinking, and unless that is fixed, no technology will live up to its potential in education.

I asked her what she thought about one of the most troubling changes that the Internet has brought about. In a world where everyone can be a publisher, the responsibility for determining the accuracy of information has shifted from the publisher to the student. When I was in school, we used encyclopedias for research. We and our teachers were reasonably certain that someone tried to check most of the facts in them. But when my daughter's generation goes to the Internet and puts a term into a search engine, they have to look at who the authors are and what their biases might be. In other words, instead of introducing them to the right authorities to turn to, we're telling them to learn to think critically. And many parents, I think probably most parents, don't want their kids to think critically. They don't want them to question authority, they don't want them to question their parents, they don't want them to question God, America, and apple pie.

Brenda replied:

> I suspect that many parents don't know how to think critically themselves and therefore they can't imagine how to transmit those skills to their children, and it scares the hell out of them. I can think of no other explanation of why the Kansas Board of education has banned the teaching of evolution. So we have this incredibly wonderful development with the Web, but not only is there no user manual, what's required is not a user manual, but a way of thinking. It's the way of thinking that gave us the Renaissance, modern science, American democracy. It involves critical thinking skills and self-reliance. Don't simply accept authority—go out and experiment for yourself.

From Expert Systems to Knowledge Management

In the earliest days of computer technology development, optimistic predictions regarding "artificial intelligence" led to expectations of AI applications (such as natural language translation and understanding) that were never achieved, even after decades of development. Although robotics, speech recognition, and other early goals of AI research have been achieved, the field has not lived up to its early hype. Much the same has happened to the AI-spinoff known as expert systems. Indeed, Avron Barr, whom I interviewed in 1983, wrote recently: "The inventors of expert systems and those of us who left the laboratory to make our fortunes commercializing the technology, had a bad case of 'inventor's myopia.' We knew exactly what the technology was good for—and we were wrong." (http://www.stanford.edu/group/scip/avsgt/expertsystems/aiexpert.html)

I talked with Barr again in 1999. He pointed out that some expert systems technology are indeed working today in standalone commercial applications:

> Financial institutions use expert systems for underwriting, credit analysis, and fraud detection, and the technology has found applications as varied as manufacturing, scheduling, and telecommunications routing. Embedded in Microsoft Office are two or three expert systems that give advice based on what is going wrong for the user. Because of the Internet, the personal computer, and knowledge management as a corporate strategy, I've come to see expert systems as a communication technology, as opposed to a problem-solving or thinking technology.

I asked him to elaborate on how expert systems, the Web, and knowledge management are coming together:

> The technology allows you to explain something to a computer system so that the computer can explain it to somebody else, later. Your knowledge can be used to help somebody while they are trying to find information, fill out a form, or operate a device. The alternative methods for sharing our experience via computer systems, which mostly involve typing text into a computer and then letting someone find and read that text, work a lot of the time but not in some important instances, and they are very inefficient and hard to keep up to date. If you are going to use text as the vehicle for moving knowledge you have to either make sure the relevant text is presented at the right moment, or depend on the user searching for text by keywords. And you have to be sure the user can understand the words you used in the text. That's the way we do it now on the web.

Expert systems add a helping hand. For example, suppose you are trying to interact with a website because you have a problem with new software you've installed. You can describe symptoms or browse through a taxonomy of different kinds of problems, but unless you are experienced in this kind of problem-solving, it is going to be a hard search. Now suppose instead that someone who knew what they were doing could listen to you and ask for some critical information and then direct you to some text that is going to solve your problem. That's where expert systems come in. They can know the key dependencies of the knowledge domain. Experts often say "it depends" when a non-expert asks a question. Expert systems lets us build dependencies into these systems in a way that can be maintained as the world changes. Improving the way people communicate by allowing the computer to understand these dependencies makes a big difference, especially now that we are all on the same computer. Helping people do work online by getting, at just the right moment, critical bits of information from someone who's been down the same road, is where the technology is really finding a home.

Toward a New Enlightenment Project

It looks like the quest to create and distribute mind-amplifying tools has been wildly successful. Now what? I'm not the only one to wonder about whether we need some new way of thinking that would enable us to use these tools to build a more humane, sustainable future for humankind. We still have wars, famines, social injustice, and epidemics and, increasingly, we are seeing that our success in building tools is threatening the viability of the biosphere.

In the seventeenth century, thinkers such as Descartes, Newton, and Bacon proposed that people died of disease and starvation in large part because we did not know how to think about the physical world. They proposed that we should invent a new way of thinking, based on reason, logic, experiment, observation, and mathematics, and by using the "new method," the human race might improve our lot. The "Enlightenment Project," as this endeavor has been called by some historians, has been remarkably successful. Science and medicine indeed improved the physical environment, and the application of reason to human affairs led to the democratic nation-states of the 18th century. Few people alive today would deny that science and rationalism have improved life for most people.

But we seem to be so much better at creating tools, selling tools, transforming our lives and societies with tools, than in understanding the best ways to use tools. It might be that such an understanding is not possible in a pluralistic world, because the fundamental questions necessary to guide our use of technology are the questions of value and meaning that scientists cannot answer, and that philosophers have been arguing about for thousands of years. Or we might be in a situation similar to that of the 17th century. We know a great deal now about how to think about the physical world, but we lack a vocabulary, a forum, a discourse for discussing how to use the knowledge we've gained. It isn't so much a question of "trying to stop progress," but of making a serious attempt to answer the question: "progress toward *what?*"

NOTES

CHAPTER TWO. THE FIRST PROGRAMMER WAS A LADY

1. B. V. Bowden, ed., *Faster Than Thought* (New York: Pitman), 15.
2. Ibid., 16.
3. Herman Goldstine, *The Computer from Pascal to von Neumann* (Princeton: Princeton University Press, 1972), 100.
4. Philip Morrison and Emily Morrison, eds., *Charles Babbage and His Calculating Engines* (New York: Dover Publications, 1961), 33.
5. Doris Langley Moore, *Ada, Countess of Lovelace: Byron's Legitimate Daughter* (New York: Harper and Row, 1977), 44.
6. Ibid., 155.
7. Morrison and Morrison, *Babbage*, 251–252.
8. Ibid., 284.
9. Bowden, *Faster Than Thought*, 18.
10. George Boole, *An Investigation of the Laws of Thought, on Which Are Founded the Mathematical Theories of Logic and Probabilities* (London: Macmillan, 1854; reprint, New York: Dover Publications, 1958), 1–3.
11. Leon E. Truesdell, *The Development of Punch Card Tabulation in the Bureau of the Census, 1890–1940* (Washington: U.S. Government Printing Office, 1965), 30–31.
12. Ibid., 31.

CHAPTER THREE. THE FIRST HACKER AND HIS IMAGINARY MACHINE

1. Alan M. Turing, "On Computable Numbers, with an Application to the Entscheidungsproblem," *Proceedings of the London Mathematical Society*, second series, vol. 42, part 3, November 12, 1936, 230–265.
2. An amusing example of an easily constructed Turing machine, using pebbles and toilet paper, is given in the third chapter of Joseph Weizenbaum, *Computer Power and Human Reason* (San Francisco: W. H. Freeman, 1976).
3. Turing, "Computable Numbers."

4. Andrew Hodges, *Alan Turing: The Enigma* (New York: Simon and Schuster, 1983), 396.
5. Ibid., 326.
6. Alan M. Turing, "Computing Machinery and Intelligence," *Mind*, vol. 59, no. 236 (1950).
7. Ibid.
8. Hodges, *Turing*, 488.

CHAPTER FOUR. JOHNNY BUILDS BOMBS AND JOHNNY BUILDS BRAINS

1. Steve J. Heims, *John von Neumann and Norbert Wiener* (Cambridge, Mass.: MIT Press, 1980), 371.
2. C. Blair, "Passing of a Great Mind," *Life*, February 25, 1957, 96.
3. Stanislaw Ulam, "John von Neumann, 1903–1957," *Bulletin of the American Mathematical Society*, vol. 64, (1958), 4.
4. Goldstine, *The Computer*, 182.
5. Daniel Bell, *The Coming of Post-Industrial Society* (New York: Basic Books, 1973), 31.
6. Katherine Fishman, *The Computer Establishment* (New York: McGraw-Hill Book Co., 1981), 22.
7. Ibid., 24.
8. Goldstine, *The Computer*, 153.
9. Ibid., 149.
10. Heims, *von Neumann and Weiner*, 186.
11. Goldstine, *The Computer*, 196.
12. Hodges, *Turing*, 288.
13. Ibid., 288.
14. Goldstine, *The Computer*, 196–197.
15. Arthur W. Burks, Herman H. Goldstine, and John von Neumann, "Preliminary Discussion of the Logical Design of an Electronic Computing Instrument," *Datamation*, September–October 1962.
16. Goldstine, *The Computer*, 242.
17. Manfred Eigen and Ruthild Winkler, *Laws of the Game* (New York: Knopf, 1981), 189, 192.

CHAPTER FIVE. EX-PRODIGIES AND ANTIAIRCRAFT GUNS

1. H. Addington Bruce, "New Ideas in Child Training," *American Magazine*, July 1911, 291–292.
2. I. Grattan-Guiness, "The Russell Archives: Some New Light on Russell's Logicism," *Annals of Science*, vol. 31 (1974), 406.
3. M. D. Fagen, ed., *A History of Engineering and Science in the Bell System: National Service in War and Peace (1925–1975)* (Murray Hill, N.J.: Bell Telephone Laboratories, Inc., 1978), 135.
4. Norbert Wiener, *Cybernetics, or Control and Communication in the Animal and the Machine* (Cambridge, Mass.: MIT Press, 1948), 8.

5. Arturo Rosenblueth, Norbert Wiener, and Julian Bigelow, "Behavior, Purpose and Teleology," *Philosophy of Science*, vol. 10 (1943), 18–24.
6. Warren McCulloch, *Embodiments of Mind* (Cambridge, Mass.: MIT Press, 1965).
7. Warren McCulloch and Walter Pitts, "A Logical Calculus of the Ideas Immanent in Nervous Activity," *Bulletin of Mathematical Biophysics*, vol. 5 (1943), 115–133.
8. Pamela McCorduck, *Machines Who Think* (San Francisco: W. H. Freeman, 1979), 66.
9. Heims, *von Neumann and Wiener*, 205.
10. Norbert Wiener, *I Am a Mathematician: The Later Life of a Prodigy* (Cambridge, Mass.: MIT Press, 1966), 325.
11. Wiener, *Cybernetics*.
12. Jeremy Campbell, *Grammatical Man* (New York: Simon and Schuster, 1982), 21.
13. Heims, *von Neumann and Wiener*, 208.
14. McCorduck, *Machines Who Think*, 42.

CHAPTER SIX. INSIDE INFORMATION

1. Claude E. Shannon, "A Symbolic Analysis of Relay and Switching Circuits," *Transactions of the AIEE*, vol. 57 (1938), 713.
2. Claude E. Shannon, "A Mathematical Theory of Information," *Bell System Technical Journal*, vol. 27 (1948), 379–423, 623–656.
3. Claude E. Shannon, "The Bandwagon," *IEEE Transactions on Information Theory*, vol. 2, no. 3 (1956), 3.
4. Noam Chomsky, *Reflections on Language* (New York: Pantheon, 1975).
5. Claude E. Shannon, "Computers and Automata," *Proceedings of the IRE*, vol. 41, 1953, 1234–1241.
6. Campbell, *Grammatical Man*, 20.

CHAPTER SEVEN. MACHINES TO THINK WITH

1. J.C.R. Licklider, "Man–Computer Symbiosis," *IRE Transactions on Human Factors in Electronics*, vol. HFE-1, March 1960, 4-11.
2. Ibid., 6.
3. Ibid.
4. Ibid., 7.
5. Ibid., 4.

CHAPTER EIGHT. WITNESS TO SOFTWARE HISTORY: THE MASCOT OF PROJECT MAC

1. Hubert Dreyfus, *What Computers Can't Do: A Critique of Artificial Reason* (New York: Harper & Row, 1972).
2. R. D. Greenblatt, D. E. Eastlake, and S. D. Crocker, "The Greenblatt Chess Program," *Conference Proceedings, American Federation of Information Processing Societies*, vol. 31 (1967), 801–810.

3. Joseph Weizenbaum, *Computer Power and Human Reason* (San Francisco: W. H. Freeman, 1976), 2–3.
4. Ibid., 116.
5. Ibid., 118–119.
6. Philip Zimbardo, "Hacker Papers," *Psychology Today*, August 1980, 63.
7. Ibid., 67–68.
8. Frank Rose, "Joy of Hacking," *Science 82*, November 1982, 66.

CHAPTER NINE. THE LONELINESS OF A LONG-DISTANCE THINKER

1. Vannevar Bush, "As We May Think," *The Atlantic Monthly*, August 1945.
2. Nilo Lindgren, "Toward the Decentralized Intellectual Workshop," *Innovation*, No. 24, September 1971.
3. Douglas C. Engelbart, "A Conceptual Framework for the Augmentation of Man's Intellect," in *Vistas in Information Handling*, vol. I, Paul William Howerton and David C. Weeks, eds. (Washington: Spartan Books, 1963), 1–29.
4. Ibid., 4–5.
5. Ibid., 5.
6. Ibid., 6–7.
7. Ibid., 14.
8. Douglas C. Engelbart, "NLS Teleconferencing Features: The Journal, and Shared-Screen Telephoning," *IEEE Digest of Papers*, CompCon, Fall 1975, 175–176.
9. Douglas C. Engelbart, "Intellectual Implications of Multi-Access Computing," *Proceedings of the Interdisciplinary Conference on Multi-Access Computer Networks*, April 1970.
10. Peter F. Drucker, *The Effective Executive* (New York: Harper & Row, 1967).
11. Peter F. Drucker, *The Age of Discontinuity: Guidelines to Our Changing Society* (New York: Harper & Row, 1968).
12. Douglas C. Engelbart, R. W. Watson, and James Norton, "The Augmented Knowledge Workshop," *AFIPS Conference Proceedings*, vol. 42 (1973), 9–21.

CHAPTER TEN. THE NEW OLD BOYS FROM THE ARPANET

1. J. C. R. Licklider, Robert Taylor, and E. Herbert, "The Computer as a Communication Device," *International Science and Technology*, April 1978.
2. Ibid., 22.
3. Ibid., 21.
4. Ibid., 27.
5. Ibid., 27.
6. Ibid., 30.
7. Ibid., 31.
8. David Canfield Smith, Charles Irby, Ralph Kimball, and Eric Harslem, "The Star User Interface: An Overview," in *Office Systems Technology* (El Segundo, Calif.: Xerox Corporation, 1982).
9. Ibid., 25.

CHAPTER ELEVEN. THE BIRTH OF THE FANTASY AMPLIFIER

1. Ted Nelson, *The Home Computer Revolution* (self-published, 1977), 120–123.
2. Michael Schrage, "Alan Kay's Magical Mystery Tour," *TWA Ambassador*, January 1984, 36.
3. Seymour Papert, *Mindstorms: Children, Computers, and Powerful Ideas* (New York: Basic Books, 1980), 183.
4. Alan Kay, "Microelectronics and the Personal Computer," *Scientific American*, September 1977, 236.
5. Alan Kay and Adele Goldberg, "Personal Dynamic Media," *Computer*, March 1977, 31.
6. Alan Kay, "Microelectronics," 236.
7. Ibid., 239.
8. Ibid., 244.
9. Ibid.
10. Ibid.
11. Ibid.

CHAPTER THIRTEEN. KNOWLEDGE ENGINEERS AND EPISTEMOLOGICAL ENTREPRENEURS

1. Avron Barr, "Artificial Intelligence: Cognition as Computation," in *The Study of Information: Interdisciplinary Messages*. Fritz Machlup and U. Mansfield, eds. (New York: John Wiley & Sons, 1983).
2. Katherine Davis Fishman, *The Computer Establishment* (New York: McGraw-Hill Book Co., 1981), 362.
3. Edward A. Feigenbaum, Bruce G. Buchanan, and Joshua Lederberg, "On Generality and Problem Solving: A Case Study Using the DENDRAL Program," in *Machine Intelligence 6*, B. Meltzer and D. Michie, eds. (New York: Elsevier, 1971), 165–190.
4. Fishman, *Computer Establishment*, 364.
5. "A Rebel in the Computer Revolution," *Science Digest*, August 1983, 96.
6. Avron Barr and Edward Feigenbaum, eds., *Handbook of Artificial Intelligence* (Los Altos, Calif.: William Kaufmann, 1981).
7. Avron Barr, J. S. Bennet, and C. W. Clancey, "Transfer of Expertise: A Theme of AI Research," Working Paper No. HPP-79-11, Stanford University, Heuristic Programming Project (1979), 1.
8. Ibid., 5.
9. Edward Feigenbaum and J. Feldman, eds., *Computers and Thought* (New York: McGraw-Hill Book Co., 1963).
10. Avron Barr, "Artificial Intelligence: Cognition as Computation," 18.
11. Ibid.
12. Ibid., p. 19.
13. Ibid., p. 22.

CHAPTER FOURTEEN. XANADU, NETWORK CULTURE, AND BEYOND

1. Ted Nelson, *Dream Machines/Computer Lib* (self-published, 1974).
2. Ted Nelson, *Literary Machines* (self-published, 1983).
3. Ibid., 1/17.
4. Ibid., 1/18.
5. Ted Nelson, "A New Home For the Mind," *Datamation*, March 1982, 174.
6. Ibid., 180.
7. Roy Amara, John Smith, Murray Turoff, and Jacques Vallee, "Computerized Conferencing, a New Medium," *Mosaic*, January–February 1976.
8. Ibid., p. 21.
9. Sarah N. Rhodes, *The Role of the National Science Foundation in the Development of the Electronic Journal"* (Washington: National Science Foundation, Division of Information Science and Technology, 1976.)

Index

About the Author

HOWARD RHEINGOLD is coauthor of *Talking Tech: A Conversational Guide to Science and Technology* (Morrow/Quill, 1983), *New Technology Coloring Book* (Bantam, 1983), and *Higher Creativity* (Tarcher, 1984). He is the editor of *The Everyone Can Build a Robot Book* (Simon & Schuster, 1984), *The Silicon Valley Guide to Financial Success in Software* (Microsoft, 1984), and *Instant Access Guide to Portable WordStar* (Bantam, 1984).

Rheingold has published articles about science, technology, and psychology in *Psychology Today*, *Esquire*, *Playboy*, and the San Francisco *Examiner & Chronicle*. He has served as an editor, writer, and scientific communications consultant to the Bank of America, Xerox Corporation, Lawrence Berkeley Laboratories, Broderbund Software, Zoetrope Studios, Paramount Pictures Corporation, and Columbia Pictures.